STRENGTH OF BEAMS, FLOORS AND ROOFS:

INCLUDING DIRECTIONS FOR

DESIGNING AND DETAILING ROOF TRUSSES,

WITH

CRITICISM OF VARIOUS FORMS
OF TIMBER CONSTRUCTION.

Prepared Especially for Carpenters and Builders.

By FRANK E. KIDDER,

Author of "The Architect's and, Builder's Pocket Book," "Building Construction and Superintendence," and "Churches and Chapels."

ILLUSTRATED WITH 164 ENGRAVINGS FROM ORIGINAL DRAWINGS.

1905

TABLE OF CONTENTS

LIST OF TABLES

▼

PREFACE

DURING the past six years, the author has contributed to *Carpentry and Building*, from time to time, series of articles bearing on the strength of wooden floors and roofs, and answered many questions pertaining to the strength of various forms of construction. These articles were so well received—the numbers containing them being long since exhausted—that it has been deemed wise to collate them in book form. This the author has endeavored to do in such a way as to make them most valuable to the student and builder and also for reference.

A few new tables have been added, and the material divided into chapters, and the tables and engravings numbered consecutively.

In the preparation of the original articles, the author tried to present the matter in the simplest possible manner consistent with accuracy, avoiding algebraic formulas and obtuse or technical language, and this object has been constantly in mind in their revision and arrangement in book form.

This work is therefore purposely very elemental, being designed for those readers who have had only a common school education; nevertheless, it is believed to be as accurate and reliable—as far as it goes—as the author's more advanced treatises. Those readers who desire more information on these subjects will find it in the author's "Architect's and Builder's Pocket Book" and in the volumes of "Building Construction and Superintendence." The third volume of the latter work will

contain examples of almost every form of trussed-roof construction and a vast amount of information relating to roof-
trusses.

For young mechanics and draughtsmen who are taking up
the study of these subjects without other assistance, this work,
however, will probably be found easier to understand and a
valuable preparation for more advanced treatises. In fact, the
author hopes that this book will be of practical value to a great
many carpenters, both young and old, and result in a more
intelligent use of building materials.

CHAPTER I.

DETERMINING THE STRENGTH OF WOODEN BEAMS.

Many persons doubtless think that the strength of wooden
beams is a matter of conjecture and not of mathematics, but except for a slight variation in the strength of the wood, due to different conditions inherent in the tree and also in the degree of
seasoning, the strength of a given beam can be very accurately
determined by simple calculations. Even with the variation due to
the wood, it is possible to determine the maximum load that it is
safe to put upon a beam, which is usually the information desired.

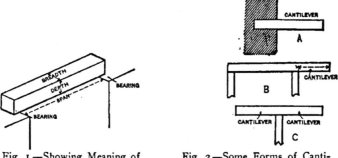

Fig. 1.—Showing Meaning of   Fig. 2.—Some Forms of Canti-
       Terms Used.   lever Beam.

Before giving any rules, however, it will be well to consider
some of the facts relating to the strength of beams. The strength
of a beam depends upon its size and shape, its span, (or if a cantilever, the projection beyond the point of support), the kind of
wood and its condition, and also the manner of loading. The following facts are also true of all rectangular wooden beams:

1

1. The strength of a beam decreases in the proportion that its span is increased. Thus the strength of a given beam, with a span of 10 feet, is one-half that of the same beam with a 5-foot span. With a span of 12 feet the strength will be five-sixths what it would be with a span of 10 feet. Or if we have a beam with a span of 20 feet and place a support under the center we just double the strength.

2. The strength of a beam increases exactly as its breadth or thickness is increased. Thus a beam 2 inches thick is twice as strong as a beam 1 inch thick, provided the other conditions remain the same.

3. The strength of a beam increases in proportion to the *square* of its depth. A 2 x 8 inch beam will be four times as strong as a 2 x 4 inch beam, and a 2 x 12 inch beam will be nine times as strong as a 2 x 4 inch beam, the square of four being 16, and of twelve 144, or nine times as great.

It follows from the second and third paragraphs that the strength of a rectangular beam is in proportion to the product of the breadth by the square of the depth if the span remains the same. A knowledge of these facts is very important for the wise use of timber.

A beam 8 x 8 contains 64 square inches in cross section, and a beam 6 x 10 contains 60 square inches, yet their strength will be in the proportion of 512 ($8 \times 8 \times 8$) to 600 ($6 \times 10 \times 10$), the 6 x 10 beam being the stronger. The strength of a 6 x 8 inch beam on edge in proportion to the strength of the same beam laid flat wise is as $6 \times 8 \times 8$ to $8 \times 6 \times 6$, or 384 to 288.

Deep beams are also very much *stiffer* than shallow beams, the resistance of a beam to bending increasing in proportion to the *cube* of the depth. The stiffness therefore of a 2 x 12 inch beam and a 2 x 10 inch beam is in the proportion of the cube of 12 to the cube of 10, or 1728 to 1000. This property of stiffness is very important in floor joists, where the span in feet is usually greater than the depth in inches, but for shorter beams it need not be considered.

In speaking of the strength or stiffness of beams the *breadth* of the beam always refers to the thickness measured horizontally, and the *depth* to the height of the beam as it sets in place, without regard to which is the larger dimension. When a beam is supported at each end the distance between supports is called the span. The distance which the ends rest on their support is called the bearing.

In the rules hereinafter given the breadth and depth of the beam are always supposed to be measured in *inches* and the span in *feet*. The meaning of the terms referred to is clearly shown in Fig. 1. Beams are also sometimes supported at three or more points, in which case they are called continuous beams. These will be considered in their proper place. There is also the *cantilever beam*, or a beam fixed at one end. The cantilever portion of the beam is that which projects beyond the support. The other end may be fixed in a wall, as at A, Fig. 2, or it may be held down by its own weight and the load on it, as at B. A beam supported at the center only, as at C, is a double cantilever, each side being considered as a cantilever. All three cases are met with in building construction, although that shown at B is the most common.

There are also different ways of loading a beam, although loads are usually classed either as distributed or concentrated. A *distributed load* is one that is applied over the entire length of the span, and when the load is uniform, as in the case of a plain brick wall of uniform height, the load is called uniformly distributed. Floor loads, although as a matter of fact not absolutely uniform, are generally considered as such. Floor joists resting on a girder may be considered as a uniformly distributed load, when the joists are not spaced more than 2 feet on centers. When they are spaced 4 feet or more on centers they should be considered as a series of concentrated loads.

A *concentrated load* is one that is applied at a single point of a beam, although in practice the "point" may be perhaps 3 feet long. An iron safe resting on the center of a beam 10 feet or more in length would be considered as a concentrated load. The end of a header framed to a trimmer is also a concentrated load, as is also a partition extending across a series of beams or joists.

The effect of a concentrated load applied at the center of a beam is just *twice as great* as if the load were uniformly distributed. When the load is applied between the center and the end the effect may be greater or less than that of a distributed load, according as the point of application is nearer to the center or to the support.

LIVE AND DEAD LOADS.

Loads are also spoken of as "live" and "dead" loads. A dead load is one that does not move of itself, such as the weight of any kind of material or a brick wall, for instance. A live load is one that is constantly moving and quickly applied. Live loads that produce a decided impact or vibrations are nearly twice as destructive as those that remain perfectly still. The principal live loads met with in building construction are moving crowds of people, particularly if they move in regular time, as in dancing or marching; machinery and wind pressure.

RULES FOR THE STRENGTH OF BEAMS.

The strength of a beam subject to almost any of the different variations of loading may be determined with about the same degree of accuracy as if simply loaded at the center, but the calculations require a considerable knowledge of mathematics, so that only a few of the more common cases can be covered by simple rules. These we will now consider.

When considering the strength of beams we usually have either one of two problems to solve—namely, to find the strength of a given beam or to determine the necessary size of beam to support a given load. The same algebraic formula really answers for both, but for the benefit of those not proficient in algebraic equations we will give a simple rule for each question, and also for each of the common conditions of support and loading. When we have to determine the strength of a given beam all of the conditions are

known, but when we wish to determine the size of beam to carry a
given load we must guess at or assume one dimension of the beam
and solve for the other. If our first guess gives a badly proportioned beam we must guess again, and do the problem over again
a second time. The quantity which represents the strength of the
wood or the resistance of the fibers to breaking is now commonly
designated as "fiber stress." In text books written previous to the
year 1885 the same quantity is called "modulus of rupture." This
quantity, of course, varies with different woods, and has been
determined by numerous experiments on beams of the different
kinds of woods. For convenience in making calculations *one-
eighteenth* of the modulus of rupture is generally used for determining the breaking strength of wooden beams, and *one-third* of
this latter value for determining the safe strength.

In the following rules this quantity will be represented by the
letter *A*, the values of this letter for the different woods used in
construction being given in Table I:

*Table I.—Values of A, Used in Determining the Safe Strength of
Beams.*

Kind of Wood.	A, in Pounds.
Chestnut	60
Hemlock	55
Oak, white	75
Pine, Georgia yellow	100
Pine, Norway	70
Pine, Oregon	90
Pine, Texas yellow	90
Pine, common white	60
Redwood	60
Spruce	70
Whitewood (poplar)	65

*To find the strength of a rectangular beam, supported at both
ends and uniformly loaded over its entire length.*

Rule 1.—Multiply twice the breadth of the beam by the square
of the depth and by the value of *A* in Table I, and divide by the
span in feet.*

*In all of these rules the breadth and depth of the beam are to be
measured in inches and the span in feet, the final result being in pounds.

To find the strength of a rectangular beam, supported at both ends and loaded at the center.

Rule 2.—Multiply the breadth of the beam by the square of the depth and by the value of *A*, Table I. Divide the product by the span in feet.

To determine the SIZE *of a rectangular beam, supported at both ends and uniformly loaded over its entire length.*

Rule 3.—Assume the depth of the beam. Multiply the span by the load, and divide by twice the square of the depth multiplied by the value of *A*. The answer will be the breadth of the beam.

Example I.—A rectangular spruce beam having a span of 16 feet is required to support a uniformly distributed load of 3780 pounds; what should be the size of the beam?

Answer.—We will assume 12 inches for the depth of the beam. The span multiplied by the load = 60,480. Twice the square of the depth multiplied by *A*, for spruce = $2 \times 12 \times 12 \times 70 = 20,160$. Divide 60,480 by 20,160 and we have 3 inches for the breadth of the beam, or a beam 3 x 12 inches will just support the load. If we assume 8 inches for the depth of the beam we shall obtain 6¾ inches for the breadth. One beam would have the same strength as the other, but the deeper beam would contain the least material and bend less.

To determine the size of a rectangular beam, supported at both ends and loaded at the center.

Rule 4.—Multiply the load by 2, and then proceed by Rule 3. That is, a load of 1000 pounds at the center will require the same size beam as a load of 2000 pounds distributed.

To determine the size of a rectangular beam, supported at both ends and carrying both a distributed load and a concentrated load at the center.

Rule 5.—Multiply the concentrated or center load by 2, and add the product to the distributed load, then proceed by Rule 3.

Example II.—A hard pine girder of 12-foot span supports a distributed load of 18,000 pounds, and also a post at the center, which sustains a load of 9600 pounds; what should be the dimensions of the girder?

Answer.—Twice the center load = 19,200 pounds. This added to the distributed load = 37,200 pounds. Assume 14 inches for the depth, and proceed by Rule 3. The product of the span by the load = 12 × 37,200 = 446,400. Twice the square of the depth multiplied by A, for hard pine = 2 × 14 × 14 × 100 = 39,200. Now 446,400 divided by 39,200 = 11⅜ inches, or it will require an 11⅜ x 14 inch girder to support both loads.

To determine what amount of concentrated load a given beam supported at both ends will safely carry at a given distance, N, *from the left support* (see Fig. 3).

Rule 6.—Multiply together the breadth, the square of the depth, the span and A, and divide the final product by four times the product of N multiplied by M, both in feet. The result will be the maximum safe load in pounds.

Example III.—A 10 x 12 inch hard pine girder, having a span of 14 feet, supports a post 4 feet from the left support; what is the maximum load that should be put on the post?

Answer.—The product of the breadth, the square of the depth, the span, and A = 10 × 144 × 14 × 100 = 2,016,000. If the span is 14 feet and N is 4 feet, M will be 10 feet. Four times the product of N by M = 160, and 2,016,000 divided by 160 = 12,600 pounds, the maximum safe load.

When the load is at the center this rule will give the same result as Rule 4.

Fig. 3.—Diagram Illustrating Rule 6.

Fig. 4.—Showing Equal Loads Concentrated at Equal Distances from Supports.

To determine the SIZE OF BEAM, supported at both ends, required to support a concentrated load applied at a given distance, from the left support.

Rule 7.—Multiply four times the load by the product of M by N, and divide the final product by the product of A times the

8   STRENGTH OF BEAMS,

square of the depth times the span. The result will be the breadth in inches.

Example IV.—What size of hard pine beam will be required to support a load of 12,600 pounds 4 feet from the left support, the span being 14 feet?

Answer.—Four times the load multiplied by the product of M by $N = 2,016,000$. Assume 12 inches for the depth; then A multiplied by the square of the depth, and the product by the span $=$ $100 \times 144 \times 14 = 201,600$, and 2,016,000 divided by 201,600 equals 10 inches, the required breadth. If we had taken 14 for the depth we would have obtained a breadth of 7 34-100 inches.

To determine the strength of a rectangular beam, loaded as in Fig. 4, M being equal to M^1 and W equal to W^1.

Rule 8.—Multiply the breadth by the square of the depth and their product by A, and divide by four times M (in feet). The result will be the safe load at each point. It should be noted that in this case the strength is not affected by the span, if we neglect the weight of the beam itself.

Example V.—What are the greatest safe loads a 10 x 12 inch hard pine beam of 12 feet span will support at a distance of 4 feet from each end?

Answer.—$10 \times 144 \times A = 144.000$, which divided by four times $M = 144,000 \div 16 = 9000$ pounds at each point.

To determine the SIZE OF BEAM required to support equal loads concentrated at equal distances from the supports, as in Fig. 4.

Rule 9.—Assume the depth: Multiply four times the load at one point by M (in feet), and divide by the square of the depth multiplied by A. The answer will be the breadth of the beam in inches.

BEAMS SUPPORTED AT BOTH ENDS, IRREGULARLY LOADED.

We have now covered all the cases of loading for which a simple rule can be given. To find the size of a beam to support several loads applied at different places, without determining the bending moment, the beam should be considered as made up of as many thicknesses as there are loads, and the thickness necessary to support each load calculated, using the same depth for each thickness; the sum of the thicknesses will be the required breadth of the beam.

Example VI.—To find the size of beam necessary to safely support the loads shown in Fig. 5, the wood being Oregon pine.

Answer.—It will be necessary to consider this as three beams, placed side by side, and each loaded as in Fig. 3. We will assume 12 inches for the depth of the beam. Then by Rule 7 the thickness required to support the load A will equal $(4 \times 2500 \times 4 \times 12) \div (A \times 144 \times 16) = 480,000 \div 207,360 = 2.31$ inches.

Thickness for load $B = (4 \times 3000 \times 7 \times 9) \div (90 \times 144 \times 16) = 756,000 \div 207,360 = 3.64$ inches.

Thickness for load $C = (4 \times 2500 \times 10 \times 6) \div (90 \times 144 \times 16) = 600,000 \div 207,360 = 2.89$ inches.

Total breadth of beam $= 2.31 + 3.64 + 2.89 = 8.84$ inches, or it will require a 9 x 12 inch beam to support the three loads.*

Example VII.—The girder shown in Fig. 6 supports a partition over its entire length and heavy floor beams, placed at equal distances of 4 feet from the supports and from each other. The parti-

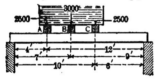

Fig. 5.—Beam Supported at Both Ends and Irregularly Loaded.   Fig. 6.—Girder Supporting Partition and Heavy Floor Beams.

tion supports a flat roof; its own weight and the weight of the roof supported by it is 16,000 pounds. The weight coming on the girder from each of the beams is 6000 pounds. The wood is long leaf Georgia pine. What should be the size of the beam?

Answer.—As the girder is symmetrically loaded, its size can be determined by two operations. The load at the center is equivalent to a distributed load of 12,000 pounds† We have then to determine the size of beam to support a distributed load of 28,000

---

*See Note, page 11.   †See Rule 4.

pounds, and the size of beam to support loads of 6000 pounds 4 feet from each support. The first should be determined by Rule 3* and the latter by Rule 9. We will assume 14 inches for the depth of the beam. The thickness of beam required to support a distributed load of 28,000 pounds is, by Rule 3, equal to $28,000 \times 16$ divided by $2 \times 14 \times 14 \times 100 = 448,000 \div 39,200 = 11\frac{1}{2}$ inches. The thickness required for the loads at A and C is, by Rule 9, equal to $4 \times 6000 \times 4 \div 14 \times 14 \times 100 = 5$ inches. The thickness required for all of the loads will be $11\frac{1}{2} + 5 = 16\frac{1}{2}$ inches, or the beam must be $16\frac{1}{2} \times 14$ inches. If it were practicable to obtain a beam 15 or 16 inches deep, it would be better to use the deeper beam. Thus if we had assumed 15 inches for the depth of our beam we would have obtained 10 inches for the thickness required for the distributed and center loads and $4\frac{1}{4}$ inches for the loads at A and C, or $14\frac{1}{4}$ inches for all of the loads.

When it is necessary to support as great loads as in the above example by wooden beams, it may be necessary to make the breadth greater than the depth, and in such a case the beam may be built up of two or more pieces of the same depth, placed side by side and bolted together. There is no objection to building up girders in this way, and in fact such a girder is often better than one made of a solid stick; but wooden girders should not be built up by spiking or bolting two or more timbers *one on top of the other*, if it is possible to obtain planks of the full depth. Compound girders made by placing one timber above another may be built so as to obtain about three-fourths of the strength of a solid timber, but it requires a particular system of keying and bolting, which is expensive, and requires careful calculation. The best method of building compound girders, and rules for computing their strength, are given in Part II of the author's work on *Building Construction and Superintendence.*

EQUIVALENT DISTRIBUTED LOAD, FOR A GIVEN CONCENTRATED LOAD.

Rule 10.—When a beam supported at both ends supports one or more concentrated loads applied at even fractions of the span from

---

*In all of these rules the breadth and depth of the beam are to be measured in inches and the span in feet, the final result being in pounds.

one support, the size of beam required to support the given loads may be most easily computed by first finding the equivalent distributed load and then finding the size of beam required to support this load by Rule 3. The equivalent distributed load for concentrated loads applied at different proportions of the span from either support may be found by multiplying the concentrated load by the corresponding factor given below:

For concentrated load at center of span, multiply by 2
For concentrated load applied at 1-3 the span, multiply by 1.78
For concentrated load applied at 1-4 the span, multiply by 1.5
For concentrated load applied at 1-5 the span, multiply by 1.28
For concentrated load applied at 1-6 the span, multiply by 1 1-9
For concentrated load applied at 1-7 the span, multiply by .98 .
For concentrated load applied at 1-8 the span, multiply by 7-8
For concentrated load applied at 1-9 the span, multiply by .79
For concentrated load applied at 1-10 the span, multiply by .72
For concentrated load applied at 1-12 the span, multiply by .61

For two equal loads applied *one-third* of the span from each support multiply one load by 2 2-3.

For two equal loads applied *one-fourth* the span from each end multiply one load by 2.

Application.—To show the application of Rule 10, we will apply it to Example VII. Here we have a distributed load of 16,000 pounds. A concentrated load at center of 6000 pounds and two concentrated loads of 6000 pounds each applied at one-fourth of the span from each support. Then by the above rule, the concentrated load at center is equal to a distributed load of 2 × 6000, or 12,000 pounds.

The two loads, 4 feet from each end (one-fourth the span), are together equal to a distributed load of 2 × 6000, or 12,000 pounds, and the total equivalent distributed load is 16,000 + 12,000 + 12,000, or 40,000 pounds.

Assuming 14 inches for the depth of the beam, the width of the beam should be, by Rule 3, $\dfrac{40,000 \times 16}{2 \times 196 \times 100} = 16\frac{1}{3}$ inches, agreeing, practically, with the value found in Example VII.

[NOTE.—The above rule (Rule 10) for finding the equivalent distributed load, and also the method used in Example VI, while absolutely correct for single loads, and also for a symmetrical application of loads, as in Fig. 6, will give an excess of strength when several concentrated loads are applied unsymmetrically as regards the span, as for instance in Figs. 5 and 7.

For a beam loaded as in Fig. 7, the equivalent distributed load found by Rule 10 would be :

$$\text{For load A, } 1000 \times 1\ 1\text{-}9 = 1111 \text{ lbs.}$$
$$\text{For load B, } 1000 \times 1.78\ \ = 1780 \text{ lbs.}$$
$$\text{For load C, } 1000 \times\ \ 2\ \ \ = 2000 \text{ lbs.}$$

Equivalent distributed load, for all three loads, 4891 pounds.

But by the correct method of bending moments, the equivalent distributed load would be but 4,000 pounds, so that Rule 10 will

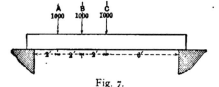

Fig. 7.

give an excess of strength of a little more than one-fifth. For Example VI, Fig. 5, the error is about 18 per cent. This error, however, is always on the safe side. The exact method is to first find the greatest bending moment produced by the load, and to then proportion the beam to the bending moment. The method of finding the bending moment for any system of loading is explained in Chapter IX of *The Architects' and Builders' Pocket Book.*]

CANTILEVER BEAMS.

To determine the maximum safe load for a beam of known dimensions, loaded and supported as in Fig. 8 or Fig. 9.

Rule 11.—To find safe load W, multiply the breadth of the beam by the square of the depth, both in inches, and this product by the value of A, and divide by $4 \times L$, in feet.

Example VIII.—What is the safe concentrated load for a spruce

---

*For the beam in Fig. 8 the distance L should always be measured from the support to a line passing through the center of the load, and in case of the beam in Fig. 9 the product of W and L must equal the product of W' and L', or if W equals W' then L must equal L'.

beam, 6 x 8 inches, fixed at one end, the point of application of the load being 6 feet from the support?

Answer.—Safe load equals $6 \times 8 \times 8 \times 70$ divided by $4 \times 6 =$ 26,880 ÷ 24 = 1120 pounds.

Fig. 8.—Cantilever Beam Supporting Load at Its Outer End.

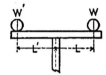

Fig. 9.—Another Form of Cantilever.

The beam will have the same strength whether loaded and sup-ported as in Fig. 8 or as in Fig. 9.

To determine the SIZE OF BEAM to support a given load applied at a fixed point from the support, as in Fig. 8 or Fig. 9.

Rule 12.—First assume the depth. To find the breadth, multi-ply four times the load, in pounds, by the distance L, in feet, and divide by the square of the depth multiplied by the value for A.

Example IX.—What size of spruce beam will be required to support a load of 1120 pounds, applied 6 feet from the support?

Answer.—Assume 8 inches for the depth of the beam. Then the breadth will be equal to $4 \times 1120 \times 6$ divided by $8 \times 8 \times 70 = 26,$-880 ÷ 4480 = 6 inches.

CANTILEVER BEAM WITH DISTRIBUTED LOAD.

To determine the maximum safe distributed load for a canti-lever beam of known dimensions.

Fig. 10.—Load Extending from
the Support.

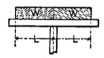

Fig. 11.—Beam Supported at the
Center.

Let W = the amount of the load, in pounds, and L the distance in feet that the load extends from the support, as in Fig. 10. If the beam is supported at the center, as in Fig. 11, W should equal the load on each side of the support.

Rule 13.—To find the safe load W, multiply the breadth by the square of the depth and the product by the value for A, and divide by two times L (in feet).

*To determine the SIZE OF BEAM required to support a known load, uniformly distributed on the beam, as in Figs. 10 and 11.**

Rule 14.—Assume the depth. To find the breadth, multiply twice the load by the distance L (in feet), and divide by the square of the depth multiplied by the value for A.

STRENGTH OF CYLINDRICAL BEAMS.

Rule 15.—To find the safe load for a cylindrical beam, as a log, first find the strength of a square beam (loaded in the same way) whose sides are equal to the diameter of the round beam, and divide the answer by 1.7. If the beam tapers slightly, as in the case of the trunk of a tree, measure the diameter at the center of the span.

Example X.—What is the safe center load for a spruce pole 12 inches in diameter at the center and with a span of 16 feet?

Answer.—By Rule 2 we find that the strength of a spruce beam, 12 inches square and 16 feet span, equals $12 \times 12 \times 12 \times 70$ divided by $16 = 7560$ pounds. Dividing this by 1.7 we have 4447 pounds for the answer.

To determine the diameter of a cylindrical beam to support a given load at the center.

Rule 16.—Multiply the span by the load and the product by 1.7 and divide by the value of A. The cube root of the result will be the answer.

Example XI.—Find the diameter of a round spruce pole of 16 feet span to support a center load of 4447 pounds.

Answer.—$4447 \times 16 \times 1.7 = 120,958.4$. Dividing this by 70, the

---

*The beam shown in Fig. 11 has the same strength as that shown in Fig. 10, provided that the distance L is equal in the two cases; *i.e.*, it makes no difference whether the cantilever end is supported by being fixed in a wall or by being balanced by an equal load on the other end of the beam.

value of *A,* we have 1728. The cube root of 1728 is 12, the required diameter of the pole.

If the load is distributed, divide it by 2 and then proceed by the above rule.

STIFFNESS OF BEAMS.

When the span of a floor or ceiling joist measured in feet exceeds the depth of the joist in inches, the beam or joist, if loaded to its full safe load, will bend more than is desirable, and often enough to crack a plastered ceiling supported by it. For this reason the size of floor joists that support plastered ceilings should be calculated by the rule for stiffness. This rule is based upon the principle of the deflection of beams, and involves a quantity known as the modulus of elasticity, which varies for different woods, and is determined by experiments upon the flexure or bending of beams under known loads. Simple rules for the stiffness of beams can only be given for the two cases of beams uniformly loaded over the entire span and of beams loaded with a concentrated load applied at the center of the span. The rules for these cases are as follows:

To determine the maximum uniformly distributed load for a rectangular beam supported at both ends that will not produce a deflection exceeding 1-30 inch per foot of span.

Rule 17.—Multiply eight times the breadth by the cube of the depth, and the product by the value for *E* (Table II), and divide by five times the square of the span.

To determine the maximum center load for a rectangular beam supported at both ends that will not produce undue deflection.

Rule 18.—Multiply the breadth by the cube of the depth, and the product by the value for *E,* and divide by the square of the span.

To determine the SIZE OF BEAM to support a given distributed load without producing undue deflection, the beam being supported at both ends.

Rule 19.—Assume the depth. Multiply five times the load by

the square of the span, and divide by eight times the cube of depth times *E*. The answer will be the breadth in inches.

To determine the SIZE OF BEAM to support a given center load without producing undue deflection, the beam being supported at both ends.

Rule 20.—Assume the depth. Multiply the load by the square of the span, and divide by the cube of the depth multiplied by *E*. The answer will be the breadth in inches.

Table II.—Value of E, to be Used in Rules 17-20.

Kind of Wood.	E, in Pounds.
Chestnut	72
Hemlock	80
Oak, white...	95
Pine, Georgia yellow.................................	137
Pine, Norway..	100
Pine, Oregon..	110
Pine, Texas yellow...................................	120
Pine, common white..................................	82
Redwood	60
Spruce	100
Whitewood (poplar)..................................	95

Example XII.—What is the maximum distributed load that a 2 x 12 inch spruce beam, 16 feet span, will support without undue deflection?

Answer.—Apply Rule 17. Eight times the breadth times the cube of the depth times $E = 8 \times 2 \times 1728 \times 100 = 2,764,800$. This divided by five times the square of the span or $1280 = 2160$ pounds, the answer.

Example XIII.—A white pine floor joist of 18-foot span has to support a uniformly distributed load of 1440 pounds; what should be the size of the beam that the deflection may not be excessive?

Answer.—We will try 10 inches for the depth of the beam, and use Rule 19. Five times the load multiplied by the square of the span $= 5 \times 1440 \times 324 = 2,332,800$. Eight times the cube of the depth $\times E = 8 \times 1000 \times 82 = 656,000$, and $2,332,800 \div 656,000 = 3\frac{1}{2}$ inches, the breadth.

If we use 12 inches for the depth we will have $2,332,800 \div 1,133,568$, which gives 2 inches for the breadth, showing that a

2 x 12 joist of 18-foot span has the same stiffness as one 3½ x 10 inches, although the latter beam contains nearly 50 per cent. more lumber than the former.

CONTINUOUS BEAMS.

A continuous beam is one which extends over three or more supports.

The formulas for determining the strength and stiffness of such beams are too elaborate to be reduced into simple rules, but it is worth while to know how the strength and stiffness of such beams compare with the strength and stiffness of a single span.

The *strength* of a continuous girder of two spans, Fig. 12, is the same as if the girder were cut over the support, when the load is distributed over the full length of the beam, but when the load is

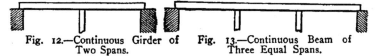

Fig. 12.—Continuous Girder of   Fig. 13.—Continuous Beam of
Two Spans.   Three Equal Spans.

applied at the center of each span the strength of the beam is increased *one-third* by making it continuous over the center support, provided that the spans are equal. If the spans are unequal the increase in the strength will be less.

The *stiffness* of a continuous beam of two equal spans is more than *doubled* by having the beam continuous over the center support, whether the load is distributed or concentrated.

In the case of a continuous beam of three equal spans, Fig. 13, the strength is increased *one-fourth* when the load is distributed, and *two-thirds* when equal loads are applied at the center of each span, and the *stiffness* is increased about 90 per cent. in both cases. It is therefore desirable to make beams continuous whenever practicable.

When beams are made continuous over three or more supports the points of greatest strain are those which come over the center supports or support, hence the beam should not be cut into at those points.

BEARING OF BEAMS ON THE WALL OR SUPPORT.

The transverse strength of a beam is not affected by the distance that the end of the beam extends onto the support, but the bearing must be sufficient that the beam will not pull off from the support when it is loaded, or that the bottom fibers of the beam will not be crushed by the load. This latter consideration is one which should be considered in the case of short beams loaded to their full capacity. Every wooden beam should have a bearing area—that is, the product of the breadth of the beam by the bearing, Fig. 1—equal to the load divided by 1000 for hard pine or oak, 500 for spruce and 400 for soft pine.

Thus a 10 x 12 inch white pine beam, of 8-foot span, might be safely loaded with 21,600 pounds if the load were uniformly distributed. Then the beam should have a bearing area = 21,600 ÷ 400, or 54 square inches. As the breadth of the beam is 10 inches, the bearing should equal 54 ÷ 10, or 5.4 inches. For floor joists a bearing of 4 inches is usually ample, and for girders a bearing of six inches is usually sufficient; 4 inches, however, should be considered as the minimum bearing, unless the beams are securely tied in place.

TABLES FOR THE STRENGTH AND STIFFNESS OF WOODEN BEAMS.

Tables III and IV will be found very convenient in figuring the safe loads for beams supported at both ends, and loaded either with a distributed load or a concentrated load applied at the center.

By following the directions given, the results obtained by the use of the tables should be the same as by using the corresponding rule, and with less figuring.

When the bending or deflection of the beam is of no importance use Table III, and when excessive bending must be avoided use Table IV.

TABLE III.—STRENGTH OF HARD-PINE BEAMS.

Table of safe quiescent loads for horizontal rectangular beams of Georgia yellow pine, one inch broad, supported at both ends, load *uniformly distributed*. For *concentrated* load at center *divide by two*. For *permanent* loads (such as masonry) reduce by 10 per cent.

Depth of Beam.	Span in Feet												
	6	8	10	12	14	15	16	18	20	22	24	25	27
ins.	lbs.	lbs.	lbs.	lbs.	lbs.	lbs.	lbs.	lbs.	lbs.	lbs.	lbs.	lbs.	lbs.
6	1,200	900	720	600	514	480							
7	1,633	1,225	980	816	700	653	612						
8	2,133	1,600	1,280	1,066	914	853	800						
9	2,700	2,025	1,620	1,350	1,157	1,080	1,012	900					
10	3,333	2,500	2,000	1,666	1,428	1,333	1,250	1,111	1,000				
12	4,800	3,600	2,880	2,400	2,056	1,920	1,800	1,600	1,440				
14	6,533	4,900	3,920	3,266	2,800	2,613	2,450	2,177	1,960	1,782	1,633	1,568	1,450
15	7,500	5,633	4,500	3,750	3,214	3,000	2,816	2,500	2,250	2,045	1,875	1,800	1,666
16	8,533	6,400	5,120	4,266	3,656	3,412	3,200	2,844	2,560	2,327	2,133	2,048	1,896

For beams of any width greater than 1 inch, multiply the load in table by the width of the beam in inches. For beams of Oregon pine, use 9-10 of tabular loads; for spruce beams, 7-10; for common white pine, 3-5, and for white oak, 3-4.

To use Table III for beams that run less than the nominal dimensions. In many localities floor joists as carried in stock are more or less scant of the nominal dimensions, and for such joists a reduction in the safe load must be made to correspond to the reduction in size. For beams having the full depth multiply the load in table by the actual breadth, as 1⅝, 1¾, 2⅞, or whatever it may be. For beams ¼ inch scant in *both* dimensions the safe load may be obtained by multiplying the safe load as given in the table by the following factors:

For beams 1¾" × 5¾"	by 1.6	For beams 1¾" × 11¾"	by 1.67
2¾" × 5¾"	by 2.52	2¾" × 11¾"	by 2.63
1¾" × 6¾"	by 1⅝	1¾" × 13¾"	by 1.68
2¾" × 6¾"	by 2.55	2¾" × 13¾"	by 2.65
1¾" × 7¾"	by 1.64	1¾" × 14¾"	by 1.69
2¾" × 7¾"	by 2.58	2¾" × 14¾"	by 2.66
1¾" × 9¾"	by 1.66	1¾" × 15¾"	by 1.7
2¾" × 9¾"	by 2.61	2¾" × 15¾"	by 2.66

Example XIV.—What is the safe load for an 8 x 12 inch Georgia pine girder, of 14 feet span?

Answer.—Safe load for 1 x 12, from table = 2056 lbs. Multiplying by the breadth, 8 inches, we have 16,448 lbs. as the safe load for the girder.

Example XV.—What is the safe load for a 3 x 12 inch white pine beam of 16 feet span ¼ inch scant in both dimensions?

Answer.—The safe load for a *hard pine* beam 1 x 12, 16 feet span, is given in the table as 1800 lbs. To find safe load for 2¾ x 11¾ beam, multiply by 2.63 = 4734 lbs., which is the safe load for a 2¾ x 11¾ inch hard pine beam of 16 feet span.

The strength of a white pine beam will be 3-5ths of this, or 2840 lbs.

To use Table IV for beams that are scant of the nominal dimensions:

The loads given in Table IV apply only to beams having the full depth indicated. To obtain the load for any thickness of beam, multiply the load in the table by the exact thickness of the beam, as 1⅝, 1¾, 2⅜, or whatever it may be.

For beams scant in both dimensions the correct load may be

TABLE IV.—MAXIMUM LOADS FOR HARD-PINE BEAMS CONSISTENT WITH STIFFNESS.

Table of maximum distributed loads which can be supported by horizontal rectangular beams of Georgia yellow pine *one inch broad*, and supported at both ends, with safety and without deflecting more than one-thirtieth of an inch per foot of span.

Depth of Beam.	Span in Feet.											Depth of Beam.
	4	6	8	10	12	14	16	18	20	22	24	
ins.	lbs.	lbs.	lbs.	lbs.	lbs.	lbs.	lbs.	lbs.	lbs.	lbs.	lbs.	ins.
6	1,800	1,200	738	473	328	242	185	146	118	97	82	6
8	3,200	2,133	1,600	1,121	778	573	438	346	280	231	194	8
9	4,050	2,700	2,025	1,596	1,108	816	624	493	399	329	277	9
10	5,000	3,333	2,500	2,000	1,520	1,120	856	676	548	452	380	10
12	7,200	4,800	3,600	2,880	2,400	1,935	1,479	1,168	950	781	656	12
14	9,800	6,533	4,900	3,920	3,226	2,800	2,348	1,855	1,503	1,240	1,042	14
15	11,266	7,500	5,633	4,500	3,750	3,214	2,816	2,281	1,850	1,525	1,282	15
16	12,800	8,533	6,400	5,120	4,266	3,656	3,200	2,769	2,244	1,851	1,536	16

For beams of any width greater than 1 inch, multiply the load in table by the width of the beam in inches.
For beams of Oregon pine, use 4-5 of tabular load; for spruce beams, 5-7, and for white pine beams, 3-5.

obtained by multiplying the load given in the tables by the following factors:

For 1¾ × 5¾ by 1.5
2¾ × 5¾ by 2.5
1¾ × 7¾ by 1.6
2¾ × 7¾ by 2.5

For 1¾ × 9¾ by 1.6
2¾ × 9¾ by 2.55
1¾ × 11¾ by 1.64
2¾ × 11¾ by 2.6
1¾ × 13¾ by 1⅝
2¾ × 13¾ by 2.6

Example XVI.—What is the maximum load, consistent with stiffness, for a Georgia pine beam 3 x 14 inches, 24 feet span?

Answer.—The load in Table IV for a 1 x 14 beam is 1042 lbs. For 3 x 14 inch it will be 3 × 1042, or 3126 lbs.

Example XVII.—What is the maximum load, consistent with stiffness, for a white pine beam measuring 2¾ x 11¾ inches, having a span of 18 feet?

Answer.—For a 1 x 12 hard pine beam the load is 1168 lbs. For 2¾ x 11¾ inch hard pine beam multiply by 2.6 = 3037 lbs. For white pine, use 3-5ths of this, or 1822 lbs.

CHAPTER II.

HOW TO DETERMINE THE STRENGTH OR SAFE LOAD OF WOODEN FLOORS.

The strength of a floor evidently depends upon the strength of the joists, headers, trimmers and girders of which it is composed, and more especially on the weakest of these, in the same way that the strength of a chain is determined by the strength of its weakest link. The joists, headers, trimmers, etc., taken as single pieces, are simply wooden beams, and their strength may be computed by the rules given in Chapter I.

The application of these rules to floors, however, may not be readily apparent to every one, and in some cases it is possible to give special rules which are more convenient to use for floors, so that a few examples, showing the method of determining the strength of floors, may prove of interest to the readers of this volume.

In dealing with "the strength of floors," we have two different problems to consider: (1) to determine the strength of a floor already built, or planned, and (2) to determine the size of beams to support a given load, with a given span. In this chapter we will consider the first of these problems—that is, to determine the strength or safe load of a given floor.

DISTINCTION BETWEEN "STRENGTH" AND "SAFE LOAD."

When we speak of the strength of a beam, we generally mean the load required to break it, and in which is included the weight of the beam itself. Now in the case of a wooden beam, its own weight, compared with the weight it will support, is usually so small that it need not generally be taken into account, but the weight of a wooden floor, meaning all of the material contained between its under and upper surface, is usually a considerable item, so that a distinction must be made between "safe strength" and "safe load." In this chapter the term "safe strength" will be used to designate the maximum weight that the floor can support with safety, including the weight of all the materials used in its construction, and for forming the ceiling below.

The *safe load* of a floor is the maximum load which can be placed on top of the finished floor, or hung beneath the floor with safety, and is found by subtracting the weight of the material from the safe strength. Thus if the safe strength of a given floor is equal to 80 pounds per square foot of floor, and the materials used in the construction of the floor (and ceiling) weigh 20 pounds per square foot of floor, then the safe load will be 60 pounds per square foot. The load on a floor usually consists either of people, furniture, machinery or merchandise. The strength, or the safe load, of a floor is usually spoken of as so many pounds per square foot, as that is the only practicable unit of measurement.

WEIGHT OF WOODEN FLOOR CONSTRUCTION.

Wooden floors usually consist of beams, commonly called "joists" or "floor joists," one or two thicknesses of flooring boards, and in a finished building of a ceiling underneath the beams. In figuring the weight of $\frac{7}{8}$-inch flooring boards it will be sufficiently accurate to estimate the weight of a single thickness at 3 pounds per square foot. The joists may also be figured at 3 pounds per foot, board measure, with the exception of hard pine and oak joists, which should be figured at 4 pounds per foot, board measure. The weight of the joists must also be reduced to their equivalent weight per square foot of floor. Thus the weight of a 2 x 12 inch joist is about 6 pounds per lineal foot. If the joists are spaced 12 inches on centers, this will be equal to 6 pounds per square foot, but if the joists are 16 inches on centers, there will be but 1 lineal foot of joist to every 1 1-3 square feet, which will be equivalent to $4\frac{1}{8}$ pounds per square foot, and if they are 20 inches on centers, the weight will be equal to 3 3-5 pounds per square foot; spaced 24 inches on centers, the weight will be 3 pounds per square foot.

The weight of a lath and plaster ceiling should be taken at 10 pounds per square foot, and of a $\frac{3}{4}$-inch wood ceiling at $2\frac{1}{2}$ pounds per square foot. Corrugated iron ceiling weighs about 1 pound per square foot.

Table V will be found convenient in figuring the weight of floor joists.

Table V.—Weight of Floor Joists per Square Foot of Floor.

Size of joists in inches.	Spruce, hemlock, white pine. Spacing in inches, center to center.		Hard pine or oak. Spacing in inches, center to center.	
	12 Pounds.	16 Pounds.	12 Pounds.	16 Pounds.
2×6	3	2¼	4	3
2×8	4	3	5⅓	4
3×8	6	4½	8	6
2×10......	5	3¾	6⅔	5
3×10......	7½	5½	10	7½
2×12......	6	4½	8	6
3×12......	9	6¾	12	9
2×14......	7	5¼	9⅓	7
3×14......	10½	8½	14	10½

We will now show how the strength and safe load of a few simple forms of floor construction may be computed.

The simplest floor is that which consists of a series of parallel

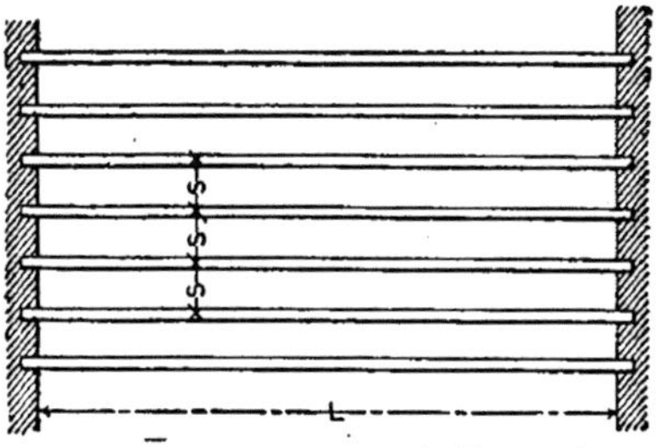

Fig. 14.—Plan of Simple Form of Floor Construction

joists of equal size, spaced a uniform distance apart, and supported at each end either by walls or partitions. Such a floor is shown in Fig. 14.

The strength of such a floor is measured by the strength of a single beam, but the strength per square foot may be obtained directly by:

Rule I.—To find the safe strength per square foot in pounds, multiply twice the breadth of a single joist by the square of the depth, and the product by the value of *A;* divide by *S* times the square of *L;* *S* being the distance between the centers of the joists *in feet,* and *L* the span in feet. *A* represents the strength of the wood, as given in Table I, page 5.

Placed in the shape of a formula the above rule will read: Safe strength per square foot in pounds $= \dfrac{2 \times B \times D^2 \times A}{S \times L^2}$ in which *×* denotes multiplication, *B* the breadth of a single joist, *D* the depth of the joist, both in inches; *S* the distance on centers, in feet, and *L* the span in feet.

Having determined the safe strength, subtract from it the weight of the floor construction, and the result will be the safe load.

Example I.—What is the safe strength per square foot of a floor formed of 2 x 10 inch spruce joists, 16 inches on centers and 16-foot span?

Answer.—Following Rule I, we multiply twice the breadth of a single joist by the square of the depth, which gives us 4 × 100 = 400, and this by the value of *A* for spruce, which gives us 28,000. *S* times the square of *L* equals 1⅓ × 256 = 341.3. Dividing 28,000 by 341.3, we have 82 pounds per square foot as the safe strength of the floor. The weight of one square foot of the floor construction, supposing that there is a plastered ceiling and double flooring, will be for the joists, 3¾ pounds; flooring, 6 pounds; lath and plaster, 10 pounds; total, 19¾ pounds. This, subtracted from 82 pounds, gives 62¼ pounds for the safe load per square foot.

Very often the joists are considerably scant of the nominal dimensions, and when such is the case the actual dimensions of the joists should be taken for the breadth and depth. Thus if the joists in the above case actually measured 1¾ x 9½ inches the safe strength would be but 64.7 pounds and the safe load about 45 pounds, or about three-fourths of what it would be were the joists of full dimensions.

STRENGTH OF FLOOR SUPPORTED BY GIRDER.

When the floor joists are supported by a girder, as in Fig. 15, the strength of the joists will be the same as if supported by a

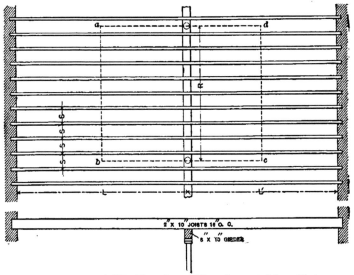

Fig. 15.—Plan and Side Elevation of Floor Supported by a Girder.

wall, but the strength of the girder must also be determined. The method of doing this is best shown by an example.

Example II.—In the floor, shown in Fig. 15, the distance L is 16 feet and the distance L' 14 feet. The distance R between the column caps supporting the girders is 12 feet. The floor joists are of 2 x 10 inch spruce, placed 16 inches on centers, and the girder 8 x 10 inch yellow pine. There will be two thicknesses of flooring and a lath and plaster ceiling below. What is the safe load of the floor?

Answer.—As the joists are of the same size, spacing and wood, as in Example I, the safe load for the 16-foot span will be the same, or 62¼ pounds. The floor area supported by the girder is that inclosed between the dotted lines *a, b, c, d,* and is equal to

$$R \times \frac{L + L^1}{2}$$ or in this case 180 square feet. The safe strength

for an 8 x 10 inch yellow pine beam, 12-foot span, we find from Table III (page 19) to be 8 × 1666, or 13,328 pounds. The safe strength of the girder per square foot of floor will be obtained by dividing the safe strength of the girder by the floor area supported. Dividing 13,328 by 180, we have 74 pounds per square foot as the *safe strength* of the girder, and subtracting 20 pounds for the weight of the floor, we have 54 pounds per square foot for the *safe load* for the floor, as measured by the strength of the girder. As the safe load for the floor joists is 62 pounds, the girder must be increased or the span shortened to utilize the full strength of the joists.*

Example III.—How shall we determine the safe load for the floor, shown in Fig. 16, all the timbers being white pine?

Answer.—By Rule 1 we find the safe strength of the common joists to equal $\dfrac{2 \times 2 \times 144 \times 60}{1\frac{1}{8} \times 324} =$ 80 pounds.

Strength of Header.—The floor area supported by the header is equal to its length multiplied by one-half of the distance *a b,* or 12 × 7 = 84 square feet. If the tail beams are framed into the header they will weaken it so as to lose, we will say, the equivalent of 1 inch of its thickness, leaving the strength of the beam about equal to that of a 5 x 12 inch. The safe distributed load for a

---

*As a general rule, if the girders have a safe strength equal to 90 per cent. of that of the floor joists it will be sufficient, as not more than ⁴⁄₁₀ of the floor area is likely to be loaded at one time.

5 x 12 inch white pine beam, 12-foot span, is by Rule 1, page 5,

$$\frac{2 \times 5 \times 144 \times 60}{12} = 7200 \text{ pounds.}$$

This, divided by the floor area it supports (84 square feet), gives about 86 pounds per square foot as the safe strength.

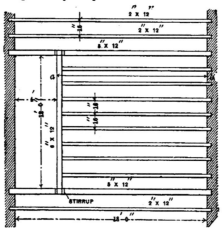

Fig. 16.—Floor with "Trimmers" and "Headers."

Strength of Trimmers.—The trimmers have to support two loads. On one side they support a floor load equal to one-half that supported by the common joists, and on the other side they support a concentrated load equal to one-half the load on the header. To support the distributed load will require a thickness of beam equal to one-half that of the common joists, or in this case 1 'inch, leaving a 4 x 12 inch beam to support the concentrated load. The safe load for a 4 x 12 inch white pine beam, 18-foot span, loaded at a point 4 feet 3 inches from one end, we find from Rule 6, page 7, to equal

$$\frac{4 \times 144 \times 18 \times 60}{4 \times 4\frac{1}{4} \times 13\frac{3}{4}} = 2662 \text{ pounds.}$$

The floor area supported by the header is 84 square feet, and as one-half will be supported at each end the floor area to be supported by the stirrup on the trimmer will be 42 square feet. Dividing the safe load, 2662 pounds, by 42, we have 63 pounds as the safe strength per square foot of floor.* Comparing now the different parts of the floor, we have:

Safe strength per square foot of common joists = 80.
Safe strength per square foot of header = 86.
Safe strength per square foot of trimmer joists = 63.

As the strength of the floor must be rated by the strength of the weakest part, we can only rate the strength of this floor at 63 pounds per square foot, and the safe load at 42½ pounds. By adding 1 inch to the thickness of the trimmer we increase its safe strength (for the concentrated load) one-fourth, making it 78¾ pounds and the safe load for the floor 58¼ pounds. If the trimmer supports stair carriages its size should be increased to offset the stair load.

Example IV.—How shall we determine the safe load per square foot of the floor, shown in Fig. 17, all of the timber being spruce, the beams covered with two thicknesses of flooring and with corrugated iron ceiling?

Answer.—This example is very much like Example III, except that the trimmers have two concentrated loads instead of one. We also assume that the trimmer *B* supports the stairs, for which an allowance of 1800 pounds must be made.

By Rule 1 we find the safe strength of the common joist equals

$$\frac{2 \times 2\frac{1}{2} \times 196 \times 70}{1\frac{1}{8} \times 484} = 106 \text{ pounds.}$$

To find the strength of the headers we allow 1 inch of the thickness for loss of strength by framing, and determine the safe distributed load of a 3 x 14 inch spruce beam, 12 feet long, by Rule 1, page 5, or Table III, to be 6860

---

*Theoretically, this method of considering a beam loaded in different ways as made up of a number of single beams or slices is not strictly correct, but the error lies on the safe side, and as the method is very much simpler than that of determining the size of beam by the bending moments, the writer feels justified in recommending it.

pounds. The floor area supported by one header is equal to
4½ x 12 feet, or 54 square feet. Dividing the safe strength of the
beam (6860 pounds) by the floor area supported (54 square feet),
we have 127 pounds as the strength of the headers per square foot
of floor.

To find the strength of trimmer A we must allow 1¼ inches of
the breadth to support the distributed load, leaving 6¼ inches to
support the headers. The safe loads for a 6¼ x 14 inch spruce

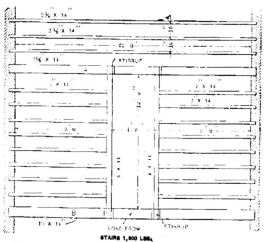

Fig. 17.—Another Example of Floor Construction.

beam, 22-foot span, loaded at points 8 feet 10 inches from each
support, we find by Rule 8, page 8, equals

$$\frac{6¼ \times 196 \times 70}{4 \times 8\,5\text{-}6} = 2427 \text{ pounds.}$$

One-half of the area supported by one of the headers is 27
square feet. Dividing 2427 by 27, we have 90 pounds as the
strength per square foot of floor.

To determine the strength of the trimmer B we must first determine how thick a beam it will require to support the 1800
pounds stair load, which is practically concentrated at the center.

Rule 4, page 6, should be used. By this rule we find that it will require a 14-inch beam, 2⅝ inches thick. Hence we must deduct 2⅝ + 1¼, or 4⅛ inches, from the breadth of our trimmer, to see how much we have left to support the headers. Deducting 4⅛ from 10 we have 5⅞ inches for thickness left to support the headers. This is a little less than we had in trimmer *A*, hence the strength will not be quite as great, but as it is not likely that the full amount of all these three loads will come on the beam at the same time, we may safely rate its strength the same as that of trimmer *A*, or 90 pounds per square foot of floor. Comparing the strength of the different portions of the floor, we find the common joists have a strength of 106 pounds, the trimmers a strength of 90 pounds, and the headers a strength of 127 pounds per square foot, showing that the trimmers are the weakest part of the construction.

The weight of the floor itself will be 6½ pounds for the joist, 6 pounds for flooring, and 1 pound for corrugated iron ceiling, or a total of 13½ pounds, making the safe load for the floor 76½ pounds each side of the stair well, and 92½ pounds elsewhere.

These four examples will serve to show the method of determining the strength and bearing capacity of a floor already constructed or planned. When the floor supports partitions, these must also be taken into account. If the partition is parallel with the beams then only the beams under the partition are affected. When the partition runs across the beams all the beams are affected and the weight of the partition, reduced to pounds per square foot of the floor area, must be added to the weight of the floor before subtracting from the safe strength to obtain the safe load.

Thus if in Example I there were a lath and plaster partition 10 feet high running across the joists at the center of the span, we first find the weight of the partition per lineal foot of the floor. A lath and plaster partition with 2 x 4 studding may be figured at 20 pounds per square foot, and as it is ten feet high the weight per lineal foot will be 200 pounds. As the load is concentrated at the center of the span it will be equivalent to a distributed load of twice that amount, or 400 pounds. As this is distributed over a span of 16 feet, dividing 400 by 16, we have 25 pounds per square foot of floor to be subtracted from the safe load already found, making the final safe load 37¼ pounds.

If the partition were 4 feet from one wall, or one-fourth of the span, the effect on the beams would be one and one-half times what it would be if the load were distributed, or 18¾ pounds per square foot. If the partition comes at a distance of one-third of the span from one support the load should be multiplied by 1.78 to obtain the equivalent distributed load. (See page 11.)

If the partition supports a floor or ceiling above, the weight on the partition should be added to the weight of the partition itself.

It will be seen from all that has gone before that the strength of a floor often depends more upon the way it is framed, the size of the headers and trimmers and the position of the partitions, than upon the strength of the common joists.

It may be well to add that while every floor should be "bridged," the bridging does not increase the bearing capacity of the floor, as a whole, but merely helps to distribute a concentrated load to three or four joists.

CHAPTER III.

HOW TO COMPUTE THE SIZE OF FLOOR TIMBERS FOR NEW BUILDINGS.

The first step in this problem will be to decide what load per square foot the floor should be calculated to support. The following values are believed by the writer to be a proper allowance for the different classes of buildings they cover—certainly they are large enough, yet on the other hand it would not, in most cases, be safe to reduce them:

Table VI.—Proper Allowance for Floor Loads per Square Foot.

	Pounds.
For dwellings, sleeping and lodging-rooms	40
For schoolrooms with fixed desks	50
For offices, upper stories	60
For stables and carriage houses	65
For banking-rooms, churches and theaters	80
For assembly halls and the corridors of all public buildings, including hotels	120
For drill-rooms	150

Floors for ordinary stores, light manufacturing and light storage should be computed for a load of not less than 120 pounds per square foot. For warehouses and heavy mercantile buildings the maximum load should be computed for each individual building. In the larger cities, the minimum safe load for which the floors of different classes of buildings shall be designed is fixed by law.

The span and arrangement of the timbers will, of course, be determined by the plans. To the floor load should be added the weight per square foot of the floor itself, which may be found from the data given in Chapter II. The sum of the load and weight of the floor will herein be designated as the *total floor load*, always in pounds per square foot.

In determining the size of the floor beams it is, perhaps, more common to use the formula for *strength* of beams, but when the ceiling below is plastered the writer recommends that the common beams be calculated by the rule for *stiffness*, as floor joists proportioned by the rule for strength will often sag so as to crack

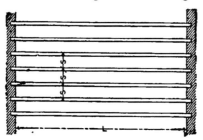

Fig. 18.—Plan of Simplest Form of Floor Construction.

a plastered ceiling, although they will not break under the load for which they have been calculated.

Common joists supported at each end, as in Fig. 18, may be computed by either of the following rules, the *depth* of the joists being first assumed:

RULE FOR STRENGTH. (A.)

To find the breadth or thickness multiply the total load per square foot by the square of the span, and this by the distance between centers of joists, in feet, and divide the product by two times the square of the depth multiplied by A. Put in the shape of a formula this becomes:

$$\text{Breadth in inches} = \frac{W \times L^2 \times S}{2 \times D^2 \times A}$$

in which W denotes the total load per square foot, L the span in feet, S the spacing on centers, also in feet; D the depth of the beam in inches and A the strength of the wood, values for which are given in Table I, page 5.

RULE FOR STIFFNESS. (B.)

To find the breadth of the beam multiply five times the total
load by the cube of the span, and this product by the distance
apart on centers in feet, and divide the product by eight times the
cube of the depth multiplied by E, or

$$\text{breadth in inches} = \frac{5 \times W \times L^3 \times S}{8 \times D^3 \times E}$$

in which E denotes the values given in Table II, page 16, for
the elasticity of the wood, and the other letters have the same
meaning as in Formula A.

RULES FOR SPACING OF FLOOR JOISTS.

It is often desirable to use a certain size of joists and space them
whatever distance apart may be necessary to obtain the desired
strength or stiffness. In such cases one of the following rules
may be used:

RULE (C.) STRENGTH.

To determine the distance from center to center of joists, in
feet, multiply twice the breadth by the square of the depth and
the product by A, and divide by the total load (per square foot)
multiplied by the square of the span, or

$$S \text{ (in feet)} = \frac{2 \times B \times D^2 \times A}{W \times L^2}$$

RULE (D.) STIFFNESS.

To determine the distance from center to center of joists, in
feet, multiply eight times the breadth by the cube of the depth
and the product by E, and divide by five times the total load multiplied by the cube of the span, or

$$S \text{ (in feet)} = \frac{8 \times B \times D^3 \times E}{5 \times W \times L^3}$$

Example I.—What should be the size of floor joists for a dwelling or lodging room, the joists to be spaced 16 inches ($1\frac{1}{3}$ feet)

on centers, with a clear span of 16 feet, double flooring to be used and the ceiling below plastered? The joists to be common white pine.

Answer.—For such a building we should allow 40 pounds per square foot for the floor load, and the floor itself will weigh about 20 pounds, making the total load 60 pounds per square foot. We will assume 10 inches for the depth of the beams. Then by Rule A, we multiply 60 by the square of 16, and this by 1⅓, which gives 20,480, and divide by two times the square of 10 multiplied by 60 (*A*), or 12,000. Making the division, we have 1.7 inches for the breadth of the beam.

Applying Rule B, we multiply five times 60 by the cube of 16 and then by 1⅓, which gives 1,638,400, and divide by eight times the cube of 10 multiplied by 82, or 656,000. Making the division, we have 2½ inches for the breadth of the beam, showing that while a 2 x 10 inch beam will be ample for strength, it will sag more than is desirable.

Example II.—Instead of finding the size of joists for the floor, described in Example I, we wish to use 2 x 10 inch joists, and will space them as may be necessary to obtain the desired strength or stiffness. What distance from centers should they be spaced?

Answer.—Following Rule C, we multiply twice the breadth (4) by the square of 10 and then by 60, which gives us 24,000, and divide by 60 multiplied by the square of 16, or 15,360. Making the division we have 1.5 feet, or 18 inches, for the distance between centers of beams. If we use Rule D, we multiply eight times 2 by the cube of 10, and the product by 82, which gives us 1,312,000, and divide by five times 60 multiplied by the cube of 16, or 1,228,800. Making the division, we have 1.06 feet, or 12¾ inches, for the distance from center to center. If we wish a good stiff floor we should not space the joists more than 13 inches on centers, but if we think our allowance for floor load is large, and are not particular if the beams sag some, we may increase the spacing to 16 inches.

When the *span* of the joists *in feet* is about equal to the *depth* in inches, the rules for strength and stiffness will agree very closely, but as the span increases, the rule for stiffness requires more lumber than the rule for strength.

Example III.—Fig. 19 shows the second floor joists of a dwelling carrying a plastered partition 9 feet high and supporting the attic joists. The weight of the attic floor and its load will be 400

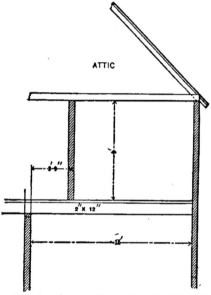

ATTIC

2″ x 12″

Fig. 19.—Section Showing Second Floor of a Dwelling where the Joists Carry a Plastered Partition.

pounds per lineal foot. The joists are to be 2 x 12 inches, *spruce,* with a single floor and plastered ceiling. What distance on centers should the joists be spaced to give ample strength?

Answer.—The first step will be to determine the total load per square foot for which the joists must be computed.

The floor joists, flooring and plastering will weigh 19 pounds per square foot if the joists are spaced 12 inches on centers, and 17½ pounds if spaced 16 inches on centers. We had better allow for 19 pounds and 40 pounds for the load per square foot on the second floor. Next we must reduce the weight of the partition and its load to an equivalent distributed load. The partition itself will weigh about 20 pounds per square foot, and as it is 9 feet high, the weight per lineal foot will be 180 pounds, which, added to the load of the partition, makes 580 pounds per lineal foot. This load is concentrated one-fourth of the span from one support, and from the table on page 11 we see that the equivalent distributed load will be obtained by multiplying the concentrated load by 1.5, which gives us 870 pounds. This distributed over a span of 15 feet is equivalent to 58 pounds per square foot. For our total distributed load on the floor beams we then have,

19 + 40 + 58, or 117 pounds per square foot.

Next we apply Rule C. Multiplying twice the breadth by the square of the depth and the product by 70 (the value of A), we have 40,320. Dividing this by the total load multiplied by the square of the span (26,325), we have 1.5 feet, or 18 inches, as the safe spacing of the floor joists. If the beams were 2 x 10 inches we would obtain a spacing of 12 inches.

As the effect of a concentrated load in producing deflection, compared with a distributed load, is not as great as the comparative breaking effects, whenever beams have a considerable concentrated load they may be *calculated by the rule for strength only*, as in the above example.

Girders, headers and trimmers, also, need only be calculated by the rules for strength, as they are usually shorter in proportion to their size, and seldom receive the full loads for which they are proportioned.

Example IV.—To determine the size of girder and floor timbers in the floor shown in Fig. 20, all the timbers being of Texas yellow pine, and the floor above being supported by posts and girders in the same way. The building is for lodging purposes, and the hight of the story is 10 feet. There is to be a double floor, and the ceilings and partitions are plastered.

Answer.—We will first determine the size and spacing of the floor joists at A, calling the span 24 feet. As Texas pine is a pretty strong wood, we will try 2 x 12 inch joists, and see how far apart they can be spaced, using the rule for stiffness (Rule D.) The weight of the floor and ceiling will be about 24 pounds per square foot, and we should allow 40 pounds for the load, making

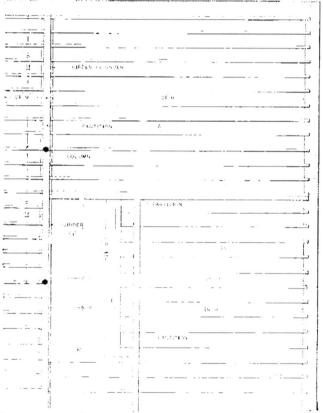

Fig. 20.—Plan of Floor of which Sizes of Girder and Timbers are to be Determined.

the total load 64 pounds. Eight times the breadth multiplied by the cube of the depth, and then by 120 (the value of E), equals 3,317,760. Five times the total load multiplied by the cube of the span equals 4,423,680. Dividing the former by the latter, we have 0.75 foot, or 9 inches, for the spacing. As this is too close for economy, we will try 2 x 14 inch joists. Making the necessary multiplications and divisions, we obtain a spacing of 1.2 feet, or 14½ inches. If we apply the rule for strength (Rule C), we obtain 1.9 feet for the safe spacing of the 2 x 14 inch joists. As 40 pounds is a good allowance for the load on the floor, we will be perfectly safe in using 2 x 14 inch joists and spacing them 16 inches on centers.

The joists at B have to support a partition, but as the span is much less, and the partition is quite near the end of the joists, it will be safe to use the same size joists at B and space them the same distance apart.

Header.—We will next consider the header H, which must be of the same depth as the floor joists. The header is 14 feet long and must support the floor half way to the wall, or a floor area of 14 x 9, or 126 square feet. As the total load per square foot is 64 pounds, this will make a floor load to be supported of 8064 pounds. Next we have the weight from the partition. The portion of the partition supported by the header is 12 feet 8 inches long, 10 feet high, and weighs 20 pounds per square foot, or 2532 pounds. Now as the partition is one-ninth of the span from the header, eight-ninths of its weight will be supported by the header and one-ninth by the wall; eight-ninths of 2532 is 2251 pounds, which added to the floor load makes a total load on the header of 10,315 pounds. Now we must determine the breadth of beam, 14 inches deep, 14-foot span, to support 10,315 pounds. This we do by Rule 3, page 6. Following the rule, we obtain 4.1 inches for the breadth of the beam. As we should allow about 1 inch for weakening by framing the tail beams, we will make the header 5 x 14 inches.

Trimmers.—We will next consider the trimmer T. This beam has four loads: a distributed floor load; a distributed load from the partition above; one-half of the load on H and a small load from the longitudinal partition.

The strip of floor supported by the trimmer will be about 12 inches wide and 24 feet long, and the load will amount to 1536 pounds. The partition above will weigh 10 × 24 × 20, or 4800 pounds.

One-half the load on H is 5158 pounds. As this is concentrated one-fourth of the span from the support, we must multiply it by 1.5 (see page 11) to obtain the equivalent distributed load, which gives 7737 pounds. About 8 inches of the longitudinal partition must be supported by the trimmer, and this will weigh 133 pounds. It is concentrated one-third of the span from the support, and we multiply by 1.78, which gives 236 pounds equivalent distributed load.

The total load for which the trimmer must be computed will therefore be:

	Pounds.
From the floor...	1,536
From the partition above............................	4,800
From the header.......................................	7,737
From the longitudinal partition......................	236
Total	14,309

Applying Rule 3, page 6, we obtain 9.8 inches for the breadth; hence the trimmer T should be 10 x 14 inches, and the header should be hung in a stirrup.

The load on the trimmer R will be the same as on trimmer T, except for the cross partition. Deducting the weight of this partition, we have 9686 pounds for the load on the beam, and following Rule 3, we find that the breadth of the beam must be 6½ inches.

Girders.—The floor area supported by girder G is equal to 12 x 24 feet, or 288 square feet. For computing the girder we can cut down our superimposed floor load to 30 pounds, which will make the total load per square foot 54 pounds and the total floor load on the girder 15,552 pounds. The girder, however, also supports a partition directly over it, which will weigh 2400 pounds, making the total load on the girder 17,952 pounds. We will assume 12 inches for the depth of the girder and compute the breadth by Rule 3. Following the rule, we obtain 8.3 inches for the breadth.

The girder G' supports a floor area at the left of 12 x 12 = 144 square feet, which represents a load of 7776 pounds. The partition over it weighs 2400 pounds, and we also have the weight from the end of the trimmer T. This load will be equal to three-fourths of the actual concentrated load from the header, two-thirds of the actual weight of the longitudinal partition (8 ins. in length) and one-half of the distributed load.

One-half of the distributed load $= 3168$ pounds, three-fourths of the actual load from the header $= 3870$ and two-thirds of the weight of the longitudinal partition $= 88$ pounds, making the total load coming from trimmer onto the girder 7126 pounds. This load is applied one-third of the span from the support, hence it must be multiplied by 1.78 to obtain the equivalent distributed load. Making the multiplication, we have 12,684 pounds.

There is also a floor area on the right of the girder, 3 x 12 feet, to be added. This will weigh 1944 pounds, and may be considered as a concentrated load acting 18 inches from the support, or one-eighth of the span. We should therefore multiply it by $\frac{7}{8}$ to obtain the equivalent distributed load, which gives 1701 pounds.

The girder must then be calculated for a distributed load of $7776 + 2400 + 12,684 + 1701 = 24,561$ pounds, which will require a beam 12 inches deep and $11\frac{1}{2}$ inches wide, but as the trimmer load was figured for a superimposed load of 40 pounds per square foot, and it is not probable that all of the floor space will be loaded at the same time, we may safely make the girder 10 x 12 inches, and it will be best to continue it the whole length of the building.

From this example it will be seen that the computations for special beams and girders consist in determining the loads which they are required to support, reducing them to an equivalent distributed load, and then computing the size of the beam by means of Rule 3, page 6. It is also very important that every specially loaded timber be computed, unless it is clearly obvious that the load on it is less than that on similar beams of same size.

STRENGTH OF FLOOR BEAMS FOR PUBLIC HALL.[*]

From S. L. C., *North Loup, Neb.*—Although I am practically a new subscriber to *Carpentry and Building*, with which I am

[*]From *Carpentry and Building*, Feb., 1904.

greatly pleased, I desire to ask the opinion of the practical readers
with regard to the carrying strength of 2 x 12 inch joist, 24-foot
span. The room is to be used as a hall and I want to know if there
is danger of overloading. Are two 2 x 6 pieces, placed one on top

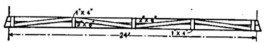

Fig. 21.—Strength of Floor Beams for Public Hall.

of the other and braced on both sides with 1 x 4 inch pieces, as
indicated in the above sketch sent herewith, stronger than 2 x 12
inch timber, and how much?

MR. KIDDER'S REPLY.

A 2 x 12 inch white pine or spruce beam is not strong enough
for the floor of a hall 24 feet wide between bearings. It is probable that 2 x 12 inch Oregon pine joists, full to dimensions, and
spaced 12 inches on centers, would be safe, but they would bend
considerably. It would be wiser to use 3 x 14's placed 16 inches
on centers.

It is impossible to build up a beam by placing one beam on top
of the other and trussing with boards so as to get the strength of
a solid timber of the same dimensions. On page 444 of Part II of
my work on "Building Construction and Superintendence" is described a method of building up two timbers to make a compound
girder, by which about 95 per cent. of the strength of the solid
beam may be obtained.

STRENGTH OF SCHOOL ROOM FLOORS.*

From C. A. D., *Penacook, N. H.*—I send herewith a drawing
showing the second-floor framing plan of a two-story wooden
school building which I am erecting. I claim that the floors are
very weak and that it will be impossible to keep a plastered ceiling
from falling off the under side of the floor timbers. The rooms

*From *Carpentry and Building* for August, 1904.

are 26 x 30 feet clear span; the floor joists, 30 in number in each room, are 2 x 10 inch spruce running across the 26-foot way and are placed 12 inches on centers. They are strengthened by three belly rod trusses, placed as shown on the plan, Fig. 22. In putting in the truss B, the timbers near the ends of the rods were bored and the rod run through the hole, as shown in Fig. 23. In the

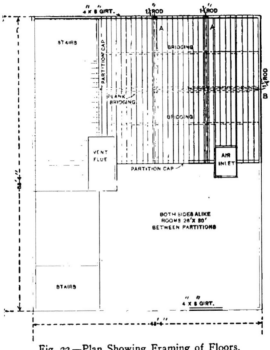

Fig. 22.—Plan Showing Framing of Floors.

center of the building notches were sawed in the under side of the floor timbers to allow the truss rod A to pass under the rod B, as shown in Figs. 23 and 24, and at the same time be high enough to clear the lath and plaster. If it were possible I would like to have the plan referred to Mr. Kidder for his opinion of the floor and trusses and have his reply published in the Correspondence Department of the paper.

MR. KIDDER'S REPLY.

I agree with "C. A. D." that the floors are very weak—so much so that I think the rooms should not be used for school purposes until the floors are strengthened in some way. In my opinion, it would have been better to have placed all four rods as shown at *A A* of Figs. 22, 23 and 24, but even then I think the floor would have been shaky.

The better method would have been to have used 3 x 14 inch spruce joists, placed 14 or 15 inches on centers and cross furred

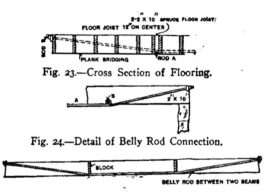

Fig. 23.—Cross Section of Flooring.

Fig. 24.—Detail of Belly Rod Connection.

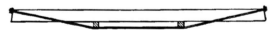

Fig. 25.—Showing Bottom of Floor Beam Strengthened at the Expense of the Top.

Fig. 26.—Form of Construction Suggested by Mr. Kidder.

the under side. The timber cannot be materially strengthened by belly rods which do not drop below the beam for the reason that wood is about equally strong in tension and compression, and if the bottom of the beam is strengthened, as in Fig. 25, the top will give way by crushing. It is true that the flooring strengthens the top of the beam to some extent and that a floor trussed, as in Fig. 25, will be stiffer than if no rods were used and also some stronger, but just how much stronger it is impossible to say without actual tests in every case. When the rod drops below the beam, as in Fig. 26, a truss is produced, which is much stronger than the single beam and the increased strength can be calculated with reasonable accuracy. See Part II of "Building Construction and Superintendence," page 450.

[NOTE.—On receipt of Mr. Kidder's reply the construction described by "C. A. D." was taken out, and 3 x 14 inch joists substituted.]

CHAPTER IV.

TYPES OF WOODEN ROOF TRUSSES, SHOWING NUMBER AND CORRECT POSITION OF MEMBERS AND THE ACTION OF THE STRESSES.

I, TRIANGULAR AND QUEEN TRUSSES; II, TRUSSES FOR FLAT ROOFS; III, SCISSORS TRUSSES.

Trusses have been used for the support of roofs for many centuries, but it is only within the last fifty years that the principles upon which they act have been reduced to a science and made so plain that any one can master them. Undoubtedly many of the trusses built to-day are copied from existing trusses, and the sizes of the pieces of which the truss is composed are determined by guessing pure and simple, although the person designing the truss would probably claim that he was guided by "experience." If it were always possible to find a truss that had the same span, was built of the same materials and loaded in exactly the same way as the truss we wished to construct, we might safely copy such a truss, provided it had stood long enough to demonstrate its safety. But it is seldom possible to find such an example to follow, and even then we might be using more material than the loads required.

It is therefore necessary to be able to design a truss for any conditions of span, loading or materials, so that we shall know with absolute certainty that it will safely carry the portion of the building supported by it and any pressure of snow or wind that may come upon the roof (with the possible exception of a cyclone), and that without using an unnecessary amount of material. To do this is not a very difficult matter, at least for the ordinary forms of wooden trusses, but it requires some study and practice, and cannot be learned from a thumb rule.

In this and the following chapters the author has endeavored to discuss the subject of wooden roof trusses in as simple a manner as possible, explaining the way in which they should be built, the action of the various parts in transmitting the strains, the way in which the size of the pieces should be determined, and how they should be joined.

DEFINITIONS.

In any discussion of trusses it is necessary to use certain terms, the meaning of which, as used by the author, must be clearly understood by the reader. In these chapters the term "truss" will be used to designate any frame work supported only at the ends (except in the case of a cantilever truss, which is supported near the center), and so designed that it cannot suffer distortion—that is, break away—without either crushing or pulling apart one of the members of which it is composed, and will exert only a *vertical* pressure on the walls or supports.

A *roof truss* is a form of truss designed for the especial purpose of supporting a roof, although it may also support the ceiling below, and perhaps a gallery or one or more floors.

By *wooden trusses* is meant trusses built principally of wood, but having iron or steel rods for some of the tension members, the term being used in distinction from trusses built entirely of iron or steel.

A *member* of a truss is any straight or curved piece of wood or iron which connects two adjacent joints of a truss, and which is essential to the stability of the truss. The term "piece" will also frequently be used to designate a particular member.

Every member of a true truss acts either as a strut or a tie.

A *tie* is a piece of material that is subject to tension—that is, a pulling strain.

A *strut* is a piece of material that is subject to a compressive force. In wooden trusses the struts are always made of wood, but the ties may be of either wood, iron or steel, the main horizontal tie being generally of wood and the vertical ties of iron. Struts must have sufficient stiffness to prevent their bending under the compression, but ties may be flexible, as a rod or cable.

The points where two or more members of the truss come together will be designated as *joints,* whether the members end at that point or not.

Purlins are horizontal beams, sometimes trussed, extended from truss to truss to support the rafters or ceiling joists.

SIMPLE FORMS OF TRIANGULAR AND QUEEN TRUSSES.

The simplest truss is that shown in Fig. 27. Such a truss, however, is only suitable for supporting a load either at the apex or on

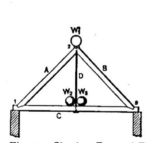
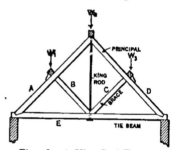

Fig. 27.—Simplest Form of Truss.   Fig. 28.—A King Rod Truss.

Fig. 29—Showing how Truss Supports Purlins and Rafters.

the tie beam. For a roof economy requires that the rafters shall not have a greater unsupported length than 12 feet, and in order to secure this it is usually necessary to introduce braces on each side of the truss, as shown in Fig. 28. These braces support the purlins, which in turn support the rafters, as shown in Fig. 29.

When the span becomes greater than 36 to 40 feet two braces are inserted on each side of the truss, as shown in Fig. 30, or the top of the truss may be cut off and a strut beam inserted between the upper ends of the principals, as shown in Fig. 31. When the span is still greater, and it is desired to utilize the attic space, a form of truss such as that shown in Fig. 32 may be used.

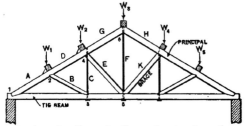

Fig. 30.—A triangular Truss for Spans Greater than 36 to 40 Feet.

The form of construction shown in Fig. 28 is commonly known as a "king rod truss," being the modern form of the old king post truss. The form of the truss represented in Fig. 30 has no generally recognized name, but may be designated as a "triangular truss." It may have two, three or even four braces on each side. The truss shown in Fig. 31 is commonly known as a "queen rod truss." Special names are given to certain members of these trusses, as indicated in the illustrations. With all of these forms of trusses it is desirable that the rafters be supported on purlins in the manner shown in Fig. 29. These purlins should extend from truss to truss, and should come directly over the ends of the braces, or as nearly so as the construction will admit.

The principles of action of any of these trusses are not varied either by the span or by the inclination of the principals, but to keep the size of the pieces within reasonable limits the inclination of the principals should not be less than 5 inches in 12, and a rise of 8 inches in 12 will generally be the most economical. The span should not be so great that the horizontal distance between the joints will exceed 12 feet, and it is generally most economical to arrange the trusses so that the distance between purlins measured on the slant of the roof will not exceed 12 feet. Ordinarily the members shown as rods should be of iron or steel, and the other pieces wooden timbers.

ACTION OF STRESSES IN TRUSSES ILLUSTRATED.

In order to know how a truss should be built one must understand the manner in which the different pieces act in supporting the loads and transmitting the stresses from one joint to another, and, finally, to the supports. It is not strictly correct to say that stresses are transmitted, as a stress is really a force acting against the end of a piece of material, and in a truss these stresses either cause a pulling or pushing at each end of the piece, according as the piece is under compression or tension. In showing how the stresses in the different members of a truss are brought into action by the loads, however, it seems clearer to speak of the stresses as passing from one member to another, and for this reason we shall so refer to them in this work.

We will first consider the action of the stresses in the truss shown in Fig. 27, taking first the stresses produced by the loads W_2 and W_3. These loads bear directly on the head or nut of the rod D, the tie beam merely acting as a bearing block at that point. The weight transmitted to the bolt head produces a pull in the rod, which carries it to the apex, where the stress is taken up and divided by the pieces A and B, half passing down A and half down B. The weight of W_1 also produces a compressive stress in A and B, the total stress in A and B being the sum of the stresses produced by W_1, W_2 and W_3. The direction of the stress in A and B is the same as the axis of the piece, and as the supports are constructed to resist a vertical force only, the horizontal component or push of the stress must be taken care of by some other piece, and it is the duty of the tie beam to take this horizontal component of the stresses in A and B.

HORIZONTAL AND VERTICAL COMPONENTS.

This leads us to a few remarks on the *components* of a force. It is a principle of mechanics that for any given force acting on a point, as, for instance, a small round ball or marble, two other forces may be substituted which, acting together on the marble,

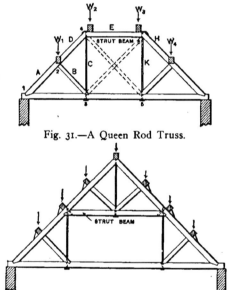

Fig. 31.—A Queen Rod Truss.

Fig. 32.—Truss Designed to Utilize Attic Space.

will have the same effect as the original single force. Thus, if we push against the ball represented in Fig. 33 with a force of 10 pounds acting in the direction of the line F (45 degrees) it will cause the ball to move forward in the direction of X. If instead of the force F we push on the ball, horizontally and vertically, with a force of 7* pounds, acting in each direction at the same time, the

*To be exact, we should say 7.07 lbs., or the $\sqrt{50}$,

ball will move forward in the same direction and with the same velocity as when acted on by the single force F; or, in other words, the forces F_1 and F_2, acting together, have the same effect as the single force F. From this fact we may assume that any oblique force may be divided into two other forces, one of which will act vertically and one horizontally, and these forces are spoken of as the vertical and horizontal *components* of the given force.

It is evident that for the forces F_1 and F_2, Fig. 33, to have the same effect on the ball as the force F they must bear a particular relation, in magnitude, to it. The value of the components of any given force can easily be determined graphically by drawing the given force in its true direction to a scale of pounds to the inch, and then drawing from one end a horizontal line and from the other end a vertical line until the two lines intersect, as in Fig. 34. The length of the horizontal and vertical lines, measured by the

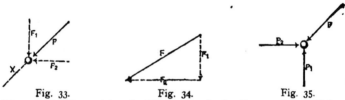

Fig. 33.　　　　　　Fig. 34.　　　　　　Fig. 35.

Diagrams for Determining the "Components" of a Force and their Value.

scale used in drawing the original force F, will give the magnitude of the components F_1 and F_2.

Admitting that two forces may be substituted for a single force, it follows that a single force may be balanced by two forces acting in a direction diametrically opposite to the components of the force and equal in magnitude to them—that is, the ball shown in Fig. 33 may be balanced by the two forces P_1 and P_2 acting as shown in Fig. 35 and each equal to 7 pounds. It should also be noticed that the sum of the components of a force is greater than the force itself, and consequently it requires more resistance to balance a single force with two forces than to balance it with one direct force. Thus, in the truss shown in Fig. 27 the combined stress in the struts A and B is greater than the sum of W_1, W_2 and W_3.

The sum of the vertical components of these stresses, however, will be just equal to the total load on the truss. With the above explanation it should be easier to follow the action of the stresses in any truss.

ACTION OF STRESSES IN FIG. 28.

We will next consider the action of the stresses in the truss shown in Fig. 28, first following the stresses produced by W_1. This load produces a pushing or compressive stress in A and B. The vertical component of the stress in B is taken up by the king

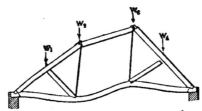

Fig. 36.—Showing Effects of a Truss Unequally Loaded.

rod, while the horizontal component is resisted by the horizontal component of the stress in C. The stress in the rod produces a compressive stress in the principals, which passes to the supports. The load W_2 is resisted directly by the two principals. If now we consider the pieces A and D, we find that they receive a stress from W_1 and W_3 direct, a stress from the pull in the king rod and also a stress from W_2. The vertical component of these three stresses is resisted by the walls and the horizontal component by the tie beam.

The stresses in the truss Fig. 30 act in the same way as in truss 28, except that there are more loads. Thus the load W_1 produces a compressive stress in both A and B. The vertical component of the stress in B is taken up by C, which carries it to joint 4, where it is resisted by D and E. The vertical component of the stress in E is taken up by the tie F, carried to joint 6, whence it passes down the principals to the supports. The horizontal component of the stress in B is resisted by the tie beam, while the horizontal components of E and F balance each other, provided the truss is symmetrically loaded. The stresses produced by W_2 and W_3 act exactly as in the truss Fig. 28.

In considering the action of the stresses in the truss shown in Fig. 31 it may be stated that the load W_1 produces a compressive stress in A and B. The vertical component of the stress in B is taken up by C and carried to joint 4, where it produces a compressive stress in D and E. The stress in D is resisted by the wall and by the tie beam, and the stress in E must be resisted by an equal stress coming from the other side of the truss. The load W_2 is resisted directly by D and E, the horizontal push in E being resisted by the corresponding push in the opposite direction produced by W_3.

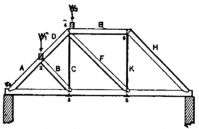

Fig. 37.—Truss Loaded on One Side Only.

That the stresses acting against the ends of E shall balance each other, it is evident that W_2 must be exactly equal to W_3, and W_1 equal to W_4. As a matter of practice, a slight inequality in the loads on opposite sides of the truss will be resisted by the stiffness of the joints, but if W_1 and W_2 are considerably greater than W_4 and W_3 then the truss will collapse, as shown in Fig. 36. To prevent this the truss should be counterbraced, as shown by the dotted lines in Fig. 31.

In order to consider the action of these counterbraces we will trace the stresses in the truss Fig. 37, which is loaded on one side only. The load W_1 produces a compressive stress in A and B. The vertical component of the stress in B is taken up by C and carried to joint 4, where it produces a compressive stress in D and F. The vertical component of the stress in F is taken up by K and

carried to joint 6, where it produces a compressive stress in E and
H. The stress in E is balanced by a portion of the horizontal component of the stress in D and the horizontal component of the
stress in H by the tie beam. In a similar manner W_2 produces a
compressive stress in D and F, a pull in K and compression in
E and H.

Now in actual construction the dead loads on a truss of this
kind, consisting of the weight of the roof and of the truss itself,
will be practically equal on each side of the truss, but one side of
the roof may be covered with snow when there is none on the
other side, and the wind can only blow on one side of the truss at
a time, consequently one side of the truss is liable frequently to be
more severely loaded than the other, and this may happen to either
side of the truss. It is also evident that an inequality in the loads
will produce the same effect as if the truss were loaded on one side
only, and as we have seen that this condition is liable to occur at
any time, it is necessary to provide for it by introducing both of
the braces shown by the dotted lines in Fig. 31.

It is, in fact, a general principle that whenever a truss is liable
to be more heavily loaded on one side than on the other all *rectangular spaces* should be braced by diagonal braces in the manner
above described. About the only exception to this rule is the truss
shown in Fig. 32, in which the rectangular space may be left open,
when there is an attic floor resting on the tie beam.

The foregoing explanation will give an idea of how the loads are
supported by simple triangular or queen rod trusses. The manner of determining the amount of the stresses, and how to proportion the members to the stresses, is explained in Chapters V
and VI.

TRUSSES FOR FLAT ROOFS.

By a "flat roof" is meant a roof having an inclination not exceeding ¾ inch to the foot, the usual inclination being about ½ inch
to the foot. Such roofs are quite commonly placed over halls,
lodge rooms, etc., and the problem of supporting such a roof is one
that may come to almost any builder.

As a general thing the space to be covered is rectangular in
shape and the roof has an inclination in only one direction, although this inclination may be either lengthways or crossways of
the roof, according to whether the gutter is to be at the end or
side of the building. If the roof is over a hall or lodge room a
finished ceiling will be required, and an air space is desirable between the roof and ceiling.

The general construction of the roof will therefore naturally
take the form shown in Figs. 38 and 39, if the inclination is lengthways of the building, and of the form shown in Fig. 40 if the roof
pitches across the building. That is, the roof and ceiling are supported by trusses placed across the building in parallel lines and

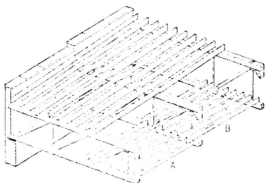

Fig. 38.—Roof where Rafters Rest on Top Chord of Truss.

from 12 to 16 feet apart. If the roof pitches, as in Figs. 38 and 39,
both the top and bottom members of the trusses will be horizontal,
and the trusses will be regulated in hight to conform to the inclination of the roof. If the inclination of the roof is across the
building then all of the trusses will be alike, and the top member
may be parallel to the roof, as in Fig. 40, or it may be level and the
purlins blocked up from it, to give the desired fall to the roof.

In supporting the roof from the trusses either of two methods
may be adopted. The more common method, probably, is to rest
the ends of the rafters directly upon the top chords of the trusses,

as shown in Fig. 38. This method answers very well for wooden roofs of moderate span, but when the span is 60 feet or over it will be more economical to support the rafters on purlins, as shown in Fig. 39. The advantages of the latter method are that the purlins, being placed over the joints of the trusses, no transverse strain is produced in the top chord, and the long side of the building is better tied to the roof. The use of purlins also permits of smaller sizes for the rafters and for the top chord, and the trusses may be spaced further apart. By bracing the purlins, as shown at A in

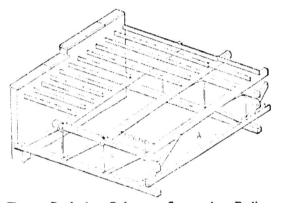

Fig. 39.—Roof where Rafters are Supported on Purlins.

Fig. 39, the trusses may be spaced from 20 to 24 feet apart. The purlins should always come either over the end of a brace or close to it.

When the rafters rest on the top chord, as in Fig. 38, the latter must be computed as a strut beam—that is, a beam which is also subject to direct compression—and as any bending effect is very weakening in a post the author considers it unwise to use this method when the distance between the vertical rods is more than 8 feet, or the truss heavily loaded.

The ceiling joists will naturally extend from truss to truss, either resting on top of the tie beams, as shown at B in Fig. 38, or framed between them, as at A. If framed between the tie beams, the bottom of the joists should be dropped $\frac{1}{2}$ inch below the bottom of the truss beams, to allow for furring the latter, unless the laths are nailed to furring strips, in which case the joists and tie beam may be flush.

If the spacing of the trusses exceeds 16 feet it will be more economical to support the ceiling joists from the tie beams of the trussed purlins. There is not so much objection to a transverse

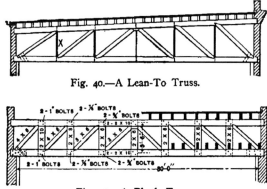

Fig. 40.—A Lean-To Truss.

Fig. 41.—A Plank Truss.

strain on the tie beams as on the top chord, because the tension in the bottom chord tends to straighten it and thus resist the bending tendency, while a compressive stress tends to increase the bending.

THE TRUSS.

For supporting the ordinary flat roof, with a ceiling below, the type of truss shown in Figs. 38 to 46 is undoubtedly the most satisfactory of any that can be devised for wooden construction, and, except in localities where iron work is very expensive, it will be as economical as any. There is no special name for this type of truss that is in general use, the term "horizontal truss" being perhaps as frequently used as any. The author prefers the name of "Howe truss," as it is essentially a Howe bridge truss without the counter

braces. For spans up to 30 feet and for light loads this truss may
be built of planks, as in Fig. 41, but it is always much better to use
rods for the vertical members, and generally there will be but a
slight difference in the cost. For heavily loaded trusses and for
spans in excess of 30 feet rods should always be used.

Howe trusses of wood with steel or wrought iron rods are practicable for spans up to 100 feet, and might be used for even greater
spans, but for a span of over 120 feet some form of arched truss
would probably be used.

For deck roofs the top chord may be inclined upward toward
the center, to conform to the shape of the roof, as shown in Fig.
42. For a deck and mansard roof the center panels should have
counter braces, as shown in Fig. 42, to resist the wind pressure
against the sides of the roof, and any unequal disposition of snow.

RULES TO BE OBSERVED WHEN DESIGNING A HOWE TRUSS.

Hight.—The hight of the truss, always measured from *center to
center* of the chords, should never be less than one-ninth of the
span, and it is not economical to make the hight more than one-
fifth of the span. As a general rule a hight of from one-seventh to
one-sixth of the span will be most economical. When the top
chord is inclined, as in Fig. 40, the hight at X—that is, at the
shortest rod—should not be less than one-ninth of the span.

Number of Panels.—A panel is the space between two adjacent
rods, or between an outer rod and the end joint (see Fig. 43).
The number of panels into which the truss is divided should be
such that the distance between any two rods shall be not more
than one and seven-tenths times the hight. Thus, if the hight is 8
feet the distance between rods should not exceed $1.7 \times 8 = 13.6$
feet. The reason for this rule is that the inclination of the braces
should not be less than about 35 degrees, as a less inclination will
produce excessive strains in the braces and joints. It is best to
keep the inclination of the braces about 45 degrees, especially if
the rafters rest on the top chord. There is no objection to making
the inclination greater than 45 degrees, except that it will usually

increase the cost. It is not material whether there be an even or odd number of panels.

Counter Braces.—If there is any chance of the truss being more heavily loaded on one end than on the other, counter braces—that is, braces in the opposite direction to that of the regular braces— should be placed in the center panels, as shown in Fig. 42. When the load is practically uniform no counter braces will be required.

Fig. 42.—A Deck Roof Truss.

Spacing of Trusses.—The most economical spacing of the trusses, all things considered, will usually be from 12 to 16 feet for spans up to 60 feet, and 14 to 20 feet for greater spans.

Spacing of Purlins.—For a roof such as is shown in Fig. 39 the most economical spacing of the purlins will be from 10 to 12 feet, center to center, as this will permit of using 6-inch rafters. As the purlins should always come over the end of a brace, the spacing of the purlins will determine the number of panels; hence, in locating the purlins the width of the panel should be considered as well as the size and span of the rafters.

When it is desired to keep the truss as low as possible, it may be an advantage to make the panels quite narrow and only have a purlin over every other joint, as in Fig. 43.

Bearing on Wall or Post.—The point where the center lines of end brace and of tie beam intersect should always come over the support, and generally at least 6 inches beyond the inner face of the wall.

LAYING OUT ROOF AND TRUSS.

The first step is to decide whether the trusses shall be placed lengthways or crossways of the building. Unless there is some special reason for doing otherwise, the trusses should be placed so as to have the shortest span.

The second step is to decide upon the number of trusses to be used, which will, of course, determine the distance between them. It will then be necessary to decide whether to support the rafters on purlins or directly on the trusses. If it is decided to use purlins the spacing of the purlins in connection with the number of panels should be decided. Last, the hight of the trusses should be determined. If it is thought best to support the rafters directly on

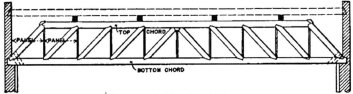

Fig. 43.—A Ten-Panel Truss.

the truss the hight of the truss should be fixed, and the number of panels made such that the braces will have an inclination of about 45 degrees, but the width of the panels should not exceed 8 feet.

Very often some special load on the roof or ceiling will fix the position of one of the rods, and, to a great extent, determine the number of panels.

RULES FOR COMPUTING THE STRESSES IN A SIX-PANEL HOWE TRUSS.

In any truss supported at the ends and having horizontal top and bottom chords, the chords resist the tension and compression due to the bending moment or transverse strain, and the rods and braces transmit the loads from the center to the supports.

Thus the load w_3, Fig. 44, is directly supported by the rod R_3, which carries it to the joint above. Here it is added to W_3, and one-half goes down the brace at the left and the other half down the brace to the right. The rod R_2 receives the load w_2 and also the vertical component* of the stress in B_3—which is equal to $\frac{1}{2} W_3 + \frac{1}{2} w_3$—and transmits the combined loads to joint 4, where

---

*The components of a stress are explained on pages 53, 54.

64   STRENGTH OF BEAMS,

the load W_2 is added to it. The sum of these loads is transmitted by brace B_2 to joint 3, and so on to joint 1.

The actual stresses in the different members of a six-panel

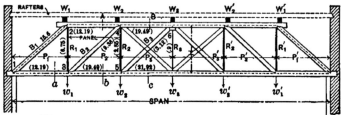

Fig. 44.—A Six-Panel Howe Truss, Showing Stresses.*

truss, as Fig. 44, may be computed by the following rules, provided that the truss is symmetrical and symmetrically loaded about a vertical line drawn through its center:

RODS.

Tension in $R_3 = w_3$ .... (1)
Tension in $R_2 = \frac{1}{2} w_3 + \frac{1}{2} W_3 + w_2$ .... (2)
Tension in $R_1 =$ tension in $R_2 + W_2 + w_1$ .... (3)

BRACES.

Compression in brace $B_1 =$ one-half the total load times length of B_1 divided by the hight H .... (4)
The total load is the sum of the loads at top and bottom.
Compression in brace $B_2 =$ stress in R_1 less w_1 times the length of B_2 divided by H .... (5)
Compression in brace $B_3 = \frac{1}{2} W_3 + \frac{1}{2} w_3$ times length of B_3 divided by H .... (6)

TIE BEAM.

Tension at $a =$ one-half the total load multiplied by the distance P_1 and divided by H .... (7)
Tension at $b =$ tension at a plus horizontal component of B_2. (8)

*Figures in ( ) denote the stresses as found in the following example.

Tension at c = tension at b plus horizontal component of B_s. (9)
The horizontal component of the stress in any brace is found
  by multiplying the stress in the brace by the width of the
  panel, and dividing by the length of the brace, measured
  from center of joints.............................(10)

COMPRESSION IN TOP CHORD.

Compression at A = tension at a. Compression at B equals
tension at b.

It will be seen from the above that the stress in the chords is
greatest at the center, as in a beam, and that the stress in the rods
and braces is least at the center and increases toward the supports.

EXAMPLE.

To illustrate the application of the above rules we will compute
the stresses in a six-panel truss of 60 feet span, P_1, P_2 and P_s being
each equal to 10 feet and with a hight, H, of 8 feet. If the trusses
are spaced 13 feet apart and support a gravel roof and plastered
ceiling the loads W_1, W_2, W_3, etc., will each be about 3 tons, and
the loads w_1, w_2, w_3 0.9 ton each. This gives all of the data necessary for determining the stresses.

We will first find the stresses in the rods.

By Rule 1 tension in $R_3 = w_3 = 0.9$ ton.

By 2, tension in $R_2 = \frac{1}{2} w_3 + \frac{1}{2} W_3 + w_2 = 0.45 + 1.5 + 0.9$
= 2.85 tons.

By 3, tension in $R_1 = 2.85 + W_2 + w_1 = 2.85 + 3 + 0.9 =$
6.75 tons.

Next find the stresses in the braces.

By Rule 4 compression in brace $B_1 = \frac{1}{2}$ the total load times
length of B_1 divided by H.

The total load = 19.5 tons and one-half of this = 9.75 tons.

The length of B_1 may be found by squaring H and P_1 and taking the square root of their sum—$H^2 = 64$; $P_1^2 = 100$. The
square root of 164 is 12.8 feet, which is the length of B_1, B_2 and
B_3 between centers of joints.

Substituting these values in the rule, we have:

$$\text{Compression in } B_1 = \frac{9.75 \times 12.8}{8} = 15.6 \text{ tons.}$$

By rule 5 compression in $B_2 = $ stress in R_1 less $w_1 = 6.75 -$

0.9, or 5.85, times length of B_2 divided by $H = \dfrac{5.85 \times 12.8}{8}$

$= 9.36$ tons.

By Rule 6 compression in $B_3 = \frac{1}{2} W_3 + \frac{1}{2} w_3$, or 1.95 times

length of B_3 divided by H, or $\dfrac{1.95 \times 12.8}{8} = 3.12$ tons.

By Rule 7 tension in tie beam at $a = \frac{1}{2}$ total load times P_1

divided by H, or $\dfrac{9.75 \times 10}{8} = 12.19$ tons.

By Rule 8 tension at $b = 12.19$ plus horizontal component of B_2.
Horizontal component of $B_2 =$ stress in B_2 times P_2 divided by

length of B_2, or $\dfrac{9.36 \times 10}{12.8} = 7.3$ tons.

Then tension at $b = 12.19 + 7.3 = 19.49$ tons.

By Rule 9 tension at $c =$ tension at $b +$ horizontal component

of $B_3 = 19.49 + \dfrac{3.12 \times 10}{12.8} = 19.49 + 2.43 = 21.92$ tons.

As the compression in the top chord at A and B is the same as the tension at a and b, we have now computed the stresses for all of the different pieces of the truss, the stresses being symmetrical each side of R_3. The stresses above obtained are shown by the numbers in parentheses in Fig. 44.

The above rules apply to any six-panel truss, provided the truss is symmetrical about the center rod. It is not necessary that P_1, P_2 and P_3 shall be equal, nor that W_1 shall be equal to W_2 or W_3, but W_1 must equal W_1^1 and P_1 must equal P_1^1.

The stresses in trusses of five or seven panels are computed in a similar manner to those in a six-panel truss, although there will be a slight difference in the formulas, due to the difference in the number of panels.

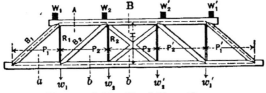

Fig. 45.—A Five-Panel Howe Truss.

RULES FOR FINDING THE STRESSES IN A FIVE-PANEL HOWE TRUSS, FIG. 45.

Tension in $R_2 = w_2$.

Tension in $R_1 = w_2 + W_2 + w_1$.

Compression in brace $B_1 = \frac{1}{2}$ total load × length of B_1 divided by H.

Compression in $B_2 = w_2 + W_2$ times length of B_2 divided by H.

Tension at $a = \frac{1}{2}$ total load times P_1 divided by H.

Tension at b = tension at a plus horizontal component of B_2.

The horizontal component of the stress in any brace is found by multiplying the stress by the panel width and dividing by the

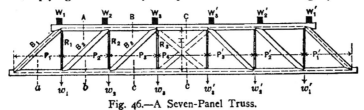

Fig. 46.—A Seven-Panel Truss.

length of the brace. Thus, if the length of B_2 is 10 feet and the distance P_2 is 8 feet, and the stress in B_2 is 5 tons, the horizontal component of the stress $= \dfrac{5 \times 8}{10} = 4$ tons.

Compression at A = tension at a.
Compression at B = tension at b.

TO FIND THE STRESSES IN A SEVEN-PANEL HOWE TRUSS, FIG. 46.

Tension in $R_3 = w_3$.
Tension in $R_2 = w_3 + W_3 + w_2$.
Tension in $R_1 = w_3 + W_3 + w_2 + W_2 + w_1$.
Compression in $B_1 = \frac{1}{2}$ total load × length of B_1 divided by H.
Compression in $B_2 =$ stress in R_1 less w_1 times length of B_2 divided by H.

Compression in $B_3 = w_3 + W_3$ times length of B_3 divided by H.
Tension at $a = \frac{1}{2}$ total load times P_1 divided by H.
Tension at $b =$ tension at a plus horizontal component of B_2.

Tension at $c =$ tension at b plus horizontal component of B_3.
(For horizontal component see explanation on page 67.)
Compression at A, B and C = tension at a, b and c, respectively.

HOW TO ESTIMATE THE ROOF AND CEILING LOADS.

The value of W_1 is found by multiplying half the sum of P_1 and P_2 by the distance between trusses, and this product by weight per square foot of the roof, including allowance for weight of truss and snow. For W_2 take one-half the sum of P_2 and P_3 and multiply as for W_1.

If panels are all equal then W_1, W_2, W_3, etc., will be equal, and each will equal the product of the panel width by the distance between trusses, multiplied by the load per square foot.*

The values for w_1, w_2, etc., are found in the same way, substituting the weight per square foot of the ceiling for the roof load.

For a lath and plaster ceiling on 2 x 6 joists the weight should be taken at 13 pounds per square foot at least, and more if there is any chance of the loft being used for storage.

If there is no ceiling to be supported the same rules may be used, omitting w_1, w_2, etc., as these would be 0.

---

*Data for estimating the weight per square foot is given in Chapter V.

When the rafters rest on the top chord the same rules apply as where purlins are used, and W_1, W_2, W_3, etc., are found in the same way, only the top chord must be computed to resist both the compressive stress and the transverse load from the rafters.

These rules apply, however, only when the truss is symmetrical and symmetrically loaded.

TABLES OF DIMENSIONS FOR HOWE TRUSSES.

For symmetrical trusses having panels of uniform width and uniformly loaded the stresses in the different parts will be proportional to the span, number of panels, hight of truss, spacing of trusses and the weight per square foot. It is therefore possible to prepare tables giving the dimensions of the parts for such trusses. The following table, computed by the author, gives the dimensions of the parts for six-panel trusses, with hights of one-sixth and one-eighth of the span, and three different spacings. These dimensions are for a flat roof of either tin, sheet iron or composition, and for a snow load of 16 pounds per square foot, which is equivalent to about 24 inches of light, dry snow; also for a lath and plaster ceiling supported by the tie beams, the chords and braces being of Norway pine and the verticals wrought iron rods.

These dimensions apply only when the rafters are supported on purlins placed at the upper joints, as in Figs. 44, 45 and 46. When the rafters rest on the top chord, as in Fig. 40, the dimensions of the latter must be greatly increased, and special calculations should be made therefor.

The dimensions given in the table may be used for trusses having a greater hight than that given, but not for trusses with a less hight, as the less the hight the greater will be the stresses.

Wherever the conditions of load, span, hight and spacing are not exactly as given in the table special calculations should be made of the stresses and corresponding dimensions, but even in such cases the table will serve somewhat as a check upon the calculations.

Table VII.—Dimensions for Six-Panel Howe Trusses.

Span	Distance Apart C to C	Total hight	Top chord	Bottom chord	Braces A	Braces B	Braces C	Rods (not upset) D	Rods (not upset) E	Rods (not upset) F
Ft.	Ft.	Ft. Ins.	Ins.	Ins.	Ins.	Ins.	Ins.	Ins.	Ins.	Ins.
36....	12.	6 7	6 x 6	6 x 8	6 x 6	6 x 4	6 x 3	1⅛	¾	⅝
		5 2	6 x 8	6 x 8	6 x 6	6 x 6	6 x 4			
	15.	6 8	6 x 8	6 x 8	6 x 6	6 x 4	6 x 3	1¼	⅞	⅝
		5 2	8 x 8	8 x 8	8 x 6	6 x 6	6 x 4			
	18.	6 8	6 x 8	6 x 8	6 x 8	6 x 6	6 x 4	1¼	⅞	⅝
		5 2	8 x 8	8 x 8	8 x 8	6 x 6	6 x 4			
42...	12.	7 7	8 x 6	8 x 8	8 x 6	8 x 4	6 x 4	1¼	⅞	⅝
		5 11	8 x 8	8 x 8	8 x 6	8 x 5	8 x 4			
	15.	7 8	8 x 8	8 x 8	8 x 6	8 x 5	6 x 4	1⅜	⅞	¾
		5 11	8 x 8	8 x 8	8 x 8	8 x 6	8 x 4			
	18.	7 8	8 x 8	8 x 8	8 x 8	8 x 6	8 x 4	1⅜	1	¾
		6 1	8 x 10	8 x 10	8 x 8	8 x 6	8 x 4			
48....	12.	8 8	8 x 8	8 x 8	8 x 8	8 x 6	8 x 4	1⅜	⅞	¾
		6 8	8 x 8	8 x 8	8 x 8	8 x 6	8 x 4			
	15.	8 8	8 x 8	8 x 8	8 x 8	8 x 6	8 x 4	1⅜	1	¾
		6 10	8 x 10	8 x 10	8 x 8	8 x 6	8 x 4			
	18.	8 8	8 x 8	8 x 8	8 x 8	8 x 6	8 x 4	1⅜	1	¾
		6 10	8 x 10	8 x 10	8 x 10	8 x 6	8 x 4			
54....	12.	9 8	8 x 8	8 x 8	8 x 8	8 x 6	8 x 4	1⅜	⅞	¾
		7 6	8 x 8	8 x 10	8 x 8	8 x 6	8 x 4			
	15.	9 8	8 x 8	8 x 8	8 x 8	8 x 6	8 x 4	1⅜	1	¾
		7 7	8 x 10	8 x 10	8 x 8	8 x 6	8 x 4			
	18.	9 10	8 x 10	8 x 10	8 x 10	8 x 8	8 x 6	1⅝	1⅛	¾
		7 7	10 x 10	10 x 10	10 x 8	8 x 8	8 x 4			
60....	12.	10 9	8 x 8	8 x 10	8 x 8	8 x 6	6 x 6	1⅜	1	⅝
		8 4	8 x 10	8 x 10	8 x 10	8 x 6	8 x 4			
	15.	10 10	8 x 10	8 x 10	8 x 10	8 x 6	6 x 6	1⅜	1⅛	¾
		8 4	10 x 10	10 x 10	10 x 8	10 x 6	8 x 4			
	18.	10 10	10 x 10	10 x 10	10 x 8	10 x 6	8 x 6	1¾	1⅛	¾
		8 4	10 x 10	10 x 10	10 x 10	10 x 6	8 x 6			
70....	12.	12 6	8 x 10	8 x 10	8 x 10	8 x 6	6 x 6	1⅜	1	¾
		9 7	10 x 10	10 x 10	10 x 8	10 x 6	8 x 6			
	15.	12 6	10 x 10	10 x 10	10 x 8	10 x 6	8 x 6	1¾	1⅛	¾
		9 9	10 x 12	10 x 12	10 x 10	10 x 8	10 x 6			
	18.	12 6	10 x 10	10 x 10	10 x 10	10 x 6	8 x 6	1⅝	1¼	⅞
		9 9	10 x 12	10 x 12	10 x 12	10 x 8	10 x 6			
80....	12.	14 2	10 x 10	10 x 10	10 x 10	10 x 6	8 x 6	1⅝	1¼	¾
		10 10	10 x 10	10 x 10	10 x 10	10 x 6	8 x 6			
	15.	14 2	10 x 10	10 x 10	10 x 10	10 x 8	8 x 6	1⅞	1¼	⅞
		11 0	10 x 12	10 x 12	10 x 10	10 x 8	10 x 6			
	18.	14 4	10 x 12	10 x 12	10 x 12	10 x 8	8 x 6	2	1⅜	1
		11 1	10 x 12	10 x 14	10 x 12	10 x 8	10 x 6			

LATTICE TRUSSES.

This form of truss was designed by Ithiel Towne, for bridges, long before iron was used in this country for such work. Several railroad bridges were built on this principle and the truss has proved very efficient in supporting loads. The truss is well adapted to the support of flat roofs in localities where large timbers and iron rods are expensive or difficult to obtain. The general shape of the truss, as used for supporting roofs, is shown in Fig. 47. The truss is composed of top and bottom chords, united by a lattice of planks and by vertical pieces at the ends. The inclination of the braces or lattice should be the same in both directions and as near 45 degrees as an even division of the span will permit. In

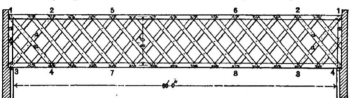

Fig. 47.—A Lattice Truss of Sixteen Spaces Built of Four 2 x 10's in 10 and 20 ft. Lengths.

the original truss the planks forming the lattice were secured to each other at their crossings, and to the chords and end pieces by wooden pins called tree-nails. In the modern truss iron bolts are used for this purpose, although dry oak pins might be used at the intersection of the braces.

The construction is very simple and can be made by any carpenter, and the materials are such as may be easily obtained in almost any village. There is no difficulty in making the truss strong enough to carry any roof load for spans up to 80 feet, but

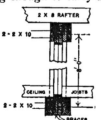

Fig. 48.—Vertical Section through Truss.

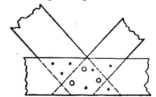

Fig. 49.—Detail of Joint 4 of Fig. 47, the Small Circles Indicating Spikes and the Large Ones 1⅛-Inch Bolts.

owing to the fact that it requires a large amount of lumber and cannot be tightened up it is not as desirable a truss to use, where iron is cheap, as the Howe truss.

A lattice truss acts in very much the same way as a beam in supporting a transverse load. The chords resist the bending moment and the bracing transmits the loads to the supports, or, in technical language, resists the shearing stress. Half of the braces are in compression and half are in tension. Uprights at the ends are necessary to receive the shear at the top and at the middle, and transmit it to the support below.

PROPORTIONS AND CONSTRUCTION.

The hight of a lattice truss, measured between the center lines of the chords, should be from one-eighth to one-sixth of the span, and the braces should be placed at an angle of about 45 degrees. When laying out a lattice truss the first step should be to determine the hight and then the number of spaces between the joints in the top and bottom chords.

To find the number of spaces multiply the span by two and divide by the hight, using the nearest whole number. Thus, if the span is 60 feet and the hight 8, there should be $\dfrac{2 \times 60}{8} = 15$ spaces. If the hight was 10 feet there should be 12 spaces. The truss shown in Fig. 47 has 16 spaces.

Having determined the hight and number of spaces, fix the center of the end joints, and divide the space between into the number of spaces determined upon, thus fixing the position of the braces. The chords should be built up of four thicknesses of planks, two on each side of the truss and breaking joint, using as long planks as can be obtained. A vertical plank should be cut between the chords, at the ends, on each side of the bracing. The braces should be bolted together where they cross and to the chords and end pieces, and also well spiked. The bottom chord should have additional bolts to tie the two thicknesses of planks together, as this member is in tension. The top chord, being in compression, will be bolted together sufficiently by the bolts at the joints and by a short bolt on each side of each butt joint.

STRESSES IN A LATTICE TRUSS.

To find the maximum stress in the chords: Multiply the total load by the span and divide by eight times the hight in feet, measured between the center lines of the chords.

The total load is obtained by multiplying the span by the distance the trusses are apart, from centers, and this product by the weight per square foot of roof and ceiling.

The weight of roof should include an allowance for the weight of the truss and also for snow load. For spans up to 60 feet lattice trusses will weigh about 5 pounds per square foot of roof surface and 6 pounds for spans from 60 to 80 feet. Ceilings will weigh about 13 pounds per square foot, and 40 pounds per square foot is generally sufficient for roof loads.

To find the stress in the end braces divide the total load by 6 and multiply by 1.4.*

The stress in the braces is greatest at the ends and decreases to nothing at the center, and hence the braces near the center of the truss may be made smaller than those at the ends.

The stress in the chords, on the contrary, is greatest at the center and decreases toward the ends, hence the center planks should be as long as can be obtained.

JOINTS.

The weakest points of a lattice truss are the joints where the braces are bolted to the top and bottom chords, particularly near the supports. These joints should have large bolts and also be well spiked. It is also a good idea to spike short pieces of planks to the sides of the end braces, cut tightly between the braces running in the opposite direction, as indicated by the dots on braces *a a* of Fig. 47.

The following table gives the dimensions for lattice trusses, built as shown in Fig. 47, for five different spans, and different

---

*This rule applies only when the braces are set at an angle closely approximating 45 degrees.

spacings and hights, which will cover nearly all of the conditions under which these trusses should be used. In localities where a fall of snow exceeding two feet in depth is liable to occur the dimensions should be increased.

TABLE VIII.—DIMENSIONS FOR LATTICE TRUSSES OF FIRST QUALITY WHITE PINE OR SPRUCE.

To support a gravel roof and plastered ceiling, allowing 20 pounds per square foot for snow.

Span	Spacing of trusses	Height out to out	No. of spaces	Size of bottom chord	Size of top chord	End braces	Inner braces	Bolts in joints 1—5 See Fig. 22
Ft.	Ft.	Ft. Ins.						Inch
40 {	12 {	5 6	16	4—2 × 6	4—2 × 6	{2 × 6	{2 × 6	{2—1
		7 2	12	4—2 × 6	4—2 × 6			
	14 {	5 7	16	4—2 × 8	4—2 × 8	{2 × 6	{2 × 6	{3— ¾
		7 3	12	4- 2 × 8	4—2 × 8			
	16 {	5 8	16	4—2 × 8	4—2 × 8	{2 × 8	{2 × 6	{3—1
		7 4	12	4—2 × 8	4—2 × 8			
50 {	12 {	6 8	16	4—2 × 8	4—2 × 8	{2 × 10	{2 × 8 and 2 × 6	{3—1
	14 {	8 8	12	4—2 × 8	4—2 × 8	{2 × 10	{2 × 6 and 2 × 8	{3—1½
		6 8	16	4—2 × 8	4—2 × 8			
	16 {	8 8	12	4—2 × 8	4—2 × 8	{2 × 10	{2 × 8 and 2 × 6	{3—1½
		6 9	16	4—2 × 8	4—2 × 10			
		8 8	12	4—2 × 8	4—2 × 8			
60 {	12 {	8 4	16	4—2 × 10	4—2 × 10	{2 × 10	{2 × 8	{3—1½
		10 10	12	4—2 × 10	4—2 × 10			
	14 {	8 4	16	4— 2 × 10	4—2 × 10	{2 × 10	{2 × 8	{3—1½
		10 10	12	4—2 × 10	4—2 × 10			
	16 {	8 4	16	4—2 × 10	4—2 × 10	{2 × 10	{2 × 8	{3—1½
		10 10	12	4—2 × 10	4—2 × 10			
70 {	14 {	9 5	16	4—2 × 10	4—2 × 12	{2 × 10	{2 × 8	{1—2¼
		12 4	12	4—2 × 10	4—2 × 12			
	16 {	9 5	16	4—2 × 10	4—2 × 12	{2 × 10	{2 × 8	{1—2¾
		12 4	12	4—2 × 10	4—2 × 10			
	18 {	9 6	16	4—2 × 12	4—2 × 12	{2 × 10	{2 × 8	{1—2½
		12 6	12	4—2 × 12	4—2 × 12			
80 {	14 {	11 0	16	4— 2× 12	4—2 × 12	{2 × 10	{2 × 8	{1—2¾
		14 0	12	4— 2× 12	4—2 × 12			
	16 {	11 2	16	4— 2× 14	4—2 × 14	{2 × 12	{2 × 10 and 2 × 8	{1—3
		14 0	12	4— 2× 12	4—2 × 12			
	18 {	11 2	16	4— 2× 14	4—2 × 14	{2 × 12	{2 × 10 and 2 × 8	{1—3
		14 1	12	4— 2× 12	4—2 × 14			

Uprights at end same size as end braces.

Referring to Fig. 47, it may be stated that each chord of the truss is built of four 2 x 10's in 10 and 20 foot lengths, the braces *a a* are 2 x 10 and the other braces 2 x 8. The joints at 1, 2, 3 and

4 have three 1½-inch bolts, the joints between 5, 6, 7 and 8 have two ⅞-inch bolts, while the other joints have two 1-inch bolts.

SCISSORS TRUSSES.

The scissors truss is a distinct type of truss much used in roofing churches and halls for the reason that it permits of a high ceiling with comparatively low walls, and is also well adapted to wooden construction. Although thousands of these trusses have been put up in all sections of the country, the writer believes that not more than half of those built even at the present time are correctly designed, and many persons seem to find it difficult to comprehend the principles of this truss and how the pieces should be put together.

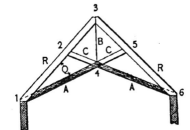

Fig. 50.—Simplest Form of Scissors Truss.

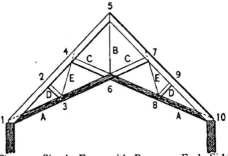

Fig. 51.—Simple Form with Brace on Each Side.

The simplest form of the scissors truss is that shown by Fig. 50. When the length between joints 1 and 2 is more than 12

feet it is desirable to introduce a brace on each side, when the
truss takes the shape shown in Fig. 51. If a level ceiling at the
center is desired, either the truss shown by Fig. 52 or that shown
by Fig. 54 will generally be the best form to use. For spans of
from 40 to 50 feet the truss shown in Fig. 55 may be used to
advantage. This shape can also be used where the roof is hipped.
Either one of the trusses shown by Figs. 52, 54, or 55 is well
adapted to the support of elliptical ceilings.

In all of these trusses the parts shown by single lines, and the
shaded or etched portions, are in tension, the other parts being
in compression. The single line members should preferably be
rods, although in light trusses boards or planks may be used instead. When the members and joints are properly proportioned,
none of the trusses shown will exert a horizontal thrust on the
walls.

PRINCIPLE OF THE SCISSORS TRUSS.

The manner in which the strains act in a truss of the type
shown by Fig. 50 is perhaps best illustrated by Figs. 56 and 57.
The loads tend to push the rafters out at the bottom or to cause
the truss to spread. This tendency to spread is resisted by the
rod b and the tie a-a, which may be considered as a flexible rod
or rope passing through a ring on the end of b.

It is evident that if b were cut in two, the rafters would imme-
diately spread until a-a became straight, as in Fig. 57, hence the
purpose of the rod b is to hold up the center of a-a. The object of
the braces c-c is to prevent the rafters from bending under the
loads W_2 and W_3. The members a and b in this truss could be
made of rope, so far as resisting the strains caused by the roof loads
is concerned, and if there were no ceiling to be supported it would
be better to use rods for the members a-a than timbers, but where
there is a ceiling to be supported, as is generally the case, it is
much cheaper and more practicable to make the tie beam of wood,
as in Fig. 50. Moreover, in actual construction the same piece
of timber generally forms the tie a and the brace c, as in Fig. 50,

because it is easier to build the truss in that way, but it should be remembered that the portion below joint 4 of Fig. 50 is in tension while the portion above is in compression.

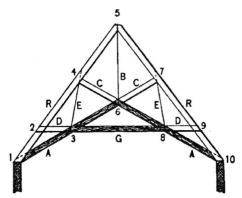

Fig. 52.—Form of Truss when Level Ceiling at Center is Wanted.

The truss Fig. 51 acts in the same way as truss 50, except that it has an additional brace and tie on each side. Whenever the brace D is inserted the tie E should also be added.

The truss 52 is the same in principle as truss 51, except that

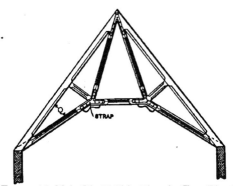

Fig. 53.—Truss with Main Tie Held in Place by Two Ties from Peak.

joints 3 and 8 are connected by a tie which reduces the strain in the center rod, and also serves as a beam to support the ceiling joists.

The truss shown in Fig. 53 differs from the first three trusses in that the main tie is held in place by two ties from the peak instead of one; for example, if the ties were of wire or rope the truss would be built as in Fig. 58. For spans of 34 or 35 feet this truss is often built entirely of wood, as in Fig. 53, the timbers being finished or cased, for ornamental effect, and secured at the joints by ornamental strap plates, bolted on each side of the timbers. When the truss is concealed the construction shown by Fig. 54 is preferable.

The truss Fig. 54 gives the same shape of ceiling as truss 52,

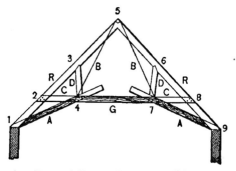

Fig. 54.—Another Form of Truss where Level Ceiling is Desired in the Center.

and when the span is greater than 36 feet it is generally the better form to use, as, there being two rods to the peak instead of one, they can be made smaller.

Fig. 59 shows a still different type of scissors truss, which is adapted to spans up to about 32 feet, but is not desirable above that limit. In this truss the cross piece c is in compression and the ties a-a are in tension their full length; for example, from 1 to 4, and from 5 to 2.

STRESSES.

No simple rules for determining the stresses in the scissors truss can be given, because scarcely two trusses have exactly the

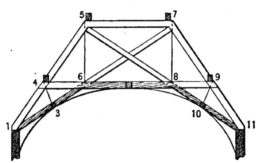

Fig. 55.—Good Form of Truss for Elliptical Building.

same proportions. The stresses in this type of truss are most easily determined by the method of graphic statics, the principles of which are explained in Chapter V.

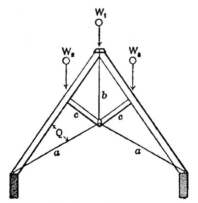

Fig. 56.—Showing how Strains Act in Truss like that in Fig. 50.

For the same span and loads the smaller the angle Q, formed
by the rafter and main tie, the greater will be the stress in all the

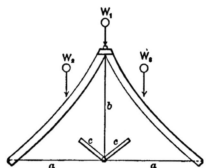

Fig. 57.—Effect Produced if Rod *b* were Cut.

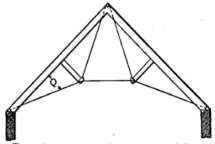

Fig. 58.—Truss in which the Ties may be of Rope or Wire.

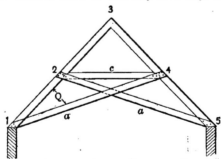

Fig. 59.—Form of Truss Adapted for Spans not Exceeding 32 Feet.

principal members; hence, when designing a scissors truss, it should be borne in mind that the steeper the rafters and the flatter the ties the less will be the strains in the truss.

The scissors truss is a dangerous truss to guess at, as the stresses in it are much greater than in a truss with a horizontal tie beam, and the strain at some of the joints, particularly that at the support, is very great.

Several good examples of scissors trusses, both rough and ornamental, are illustrated in the author's work, "Churches and Chapels."

Examples of light scissors trusses are also given in Chapter IX.

CHAPTER V.

DETERMINING THE STRESSES IN ROOF TRUSSES.

The various steps to be pursued in designing a roof truss, and by this we mean proportioning the parts and joints to the stresses that will be produced when the truss receives its greatest load, may be described as follows:

1. Laying out the roof and trusses on plan and section.
2. Computing the greatest possible loading of the truss, and its disposition.
3. Determining the resulting stress in each member of the truss.
4. Computing the size of the truss members and detailing the joints.

The first step presumes sufficient knowledge on the part of the designer to decide on the kind of truss it is best to use, its shape and hight, and the distance apart the trusses shall be spaced. These points will be determined largely by the width of the building, the inclination of the roof, and the shape of the ceiling below.

Before any computations are made a section should be drawn, showing the walls or supports, the roof and ceiling, and the approximate design of the truss. The spacing of the trusses should be determined at the same time. This drawing will enable the designer to get the necessary measurements for computing the loads at the different joints of the truss.

In drawing the truss the sizes of the pieces must be assumed, or guessed at, and after the stresses are determined they can be accurately computed, and the drawing revised.

DETERMINING ROOF LOADS.

After making the section and truss drawing, the next step will be to decide upon the loads per square foot of roof and ceiling surfaces for which the truss shall be computed. To do this one must know the construction of the roof, the kind of roofing material that will be used, the possible snow load, and the character of the ceiling, if any.

The weight of the roof will generally be made up of the following items:

Allowance for weight of truss itself.
Weight per square foot of purlins.
Weight per square foot of rafters.
Weight per square foot of sheathing.
Weight per square foot of roof covering.
Allowance per square foot for snow and wind.

For making up these weights the following data will be found of great assistance:

*TABLE IX.—Approximate Weight Per Square Foot of Roof Surface of Wooden Trusses.**

(Prepared by the Author.)

Span. Feet.	½ pitch. Pounds.	⅓ pitch. Pounds.	¼ pitch. Pounds.	Flat. Pounds.
Up to 36	3	3½	3¾	4
36 to 50	3¼	3¾	4	4½
50 to 60	3½	4	4½	4¾
60 to 70	3¾	4½	4¾	5¼
70 to 80	4¼	5	5½	6
80 to 90	5	6	6½	7
90 to 100	5¾	6¾	7	8
100 to 110	6½	7½	8	9
110 to 120	7	8½	9	10

TABLE X.—Weight of Rafters Per Square Foot of Roof Surface.

Size of Rafters.	Spruce, hemlock, white pine Spacing in inches, c. to c.			Hard pine. Spacing in inches, c. to c.		
	16	20	24	16	20	24
Inches.	Pounds.	Pounds.	Pounds.	Pounds.	Pounds.	Pounds.
2 x 4	1½	1 1-5	1	2	1 3-5	1 1-3
2 x 6	2¼	1 4-5	1½	3	2 2-5	2
2 x 7	2⅝	2 1-10	1¾	3½	2 4-5	2 1-3
2 x 8	3	2 2-5	2	4	3 1-5	2 2-3
2 x 10	3¾	3	2½	5	4	3 1-3

Wooden purlins will weigh about 2 pounds per square foot of roof surface, when the distance between trusses is from 12 to 16 feet.

Sheathing 1 inch thick will weigh about 3 pounds per square foot for the soft woods and 4 pounds for the hard woods and pitch pine.

*For scissors trusses increase one-third.

TABLE XI.—Approximate Weight Per Square Foot of Roof Surface for Roofing Materials.

Shingles, common, 2½ pounds; extra thick or long, 3 pounds.
Slates, average, 6½ pounds; thick, 9 pounds.
Plain tiles or clay shingles, 11 to 14 pounds.
Improved flat tiles, 6½ to 7 pounds.
Ludowici tile, 8 pounds.
Spanish tiles (clay), old style, two parts, 19 pounds; new style, one part, 8 pounds.
For tiles laid in mortar, add 10 pounds per square foot.
Tin roofing, sheets or shingles, including one thickness of felt, 1 pound.
Copper roofing, sheets, 1½ pounds; tiles, 1¾ pounds.
Corrugated iron, No. 26, 1 pound; No. 24, 1 1-5 pounds; No. 22, 1½ pounds.
Standing seam steel roofing, 1 pound.
Five-ply felt and gravel roof, 6 pounds.
Four-ply felt and gravel roof, 5½ pounds.
Ready roofing, in rolls, 0.6 to 1 pound.

ALLOWANCE FOR SNOW.

In making an allowance for snow, one's judgment must be exercised to a considerable degree, as the maximum snowfall varies widely in different localities, and the amount of snow that may lodge upon a roof will depend in a great measure upon the inclination of the roof and its exposure to the wind, also somewhat upon the roof covering and whether or not snow guards are used.

The weight of dry, freshly fallen snow is commonly given at 8 pounds per cubic foot, while saturated snow, or snow mixed with hail or sleet, may weigh as much as 32 pounds per cubic foot. Dry snow may attain a depth of 3 feet and possibly more in some localities, but snow weighing as much as 32 pounds per cubic foot will hardly ever be found more than 16 inches in depth, even on a flat roof.

As the maximum snow and wind cannot well be exerted on a roof both at the same time, because the wind would blow the snow off, a single allowance may be made for both kinds of loads when computing the stresses for the ordinary types of wooden trusses.

ALLOWANCE FOR WIND AND SNOW COMBINED.

In the opinion of the writer the allowances for both wind and snow given in Table XII may safely be used for ordinary conditions and without requiring an undue amount of material.*

TABLE XII.—*Allowance for Wind and Snow Combined in Pounds Per Square Foot of Roof Surface.*

Location.	Pitch of Roof.					
	60°	45°	1-3	1-4	1-5	1-6
Northwest States............	30	30	25	30	37	45
New England States........	30	30	25	25	35	40
Rocky Mountain States......	30	30	25	25	27	35
Central States..............	30	30	25	25	22	30
Southern and Pacific States..	30	30	25	25	22	20

CEILINGS.

For computing the weight of ceilings, the weight of the joists may be taken from Table X, or computed on the basis of 3 pounds per foot board measure for soft woods and 4 pounds for hard woods.

For lath and plaster allow 10 pounds per square foot. For ¾-inch ceiling 2½ pounds, and for metal ceilings with furring 1½ to 2 pounds.

The author usually adds from 3 to 5 pounds per square foot for occasional loads on the ceiling, such as persons walking or climbing over it.

COMPUTING THE TRUSS OR JOINT LOADS.

After the loads per square foot are determined upon, the loads coming on the joints of the truss should be computed.

---

*To be theoretically correct, the stresses due to wind pressure should be determined separately from those due to the vertical loads, but as the method above explained will give stresses sufficiently large to insure safety for the ordinary types of wooden trusses, without any appreciable waste of material, and as the more correct method is intricate and laborious, this method of treating wind pressure is generally pursued for ordinary types of trusses. For trusses with curved chords, or a very steep inclination, the more exact method should be followed.

Calculations for the stresses in a truss are always based on the assumption that the loads are transferred to the joints, and that the various pieces are hinged at the joints, as on a pin. When

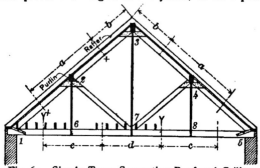

Fig. 60.—Simple Truss Supporting Roof and Ceiling.

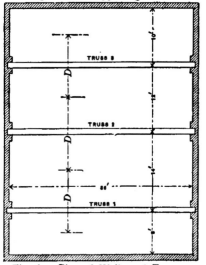

Fig. 61.—Plan of Walls and Trusses.

purlins are used to support the rafters they should always bear on the truss as near the joints as practicable.

The method of computing the joint, or panel, loads, as they

are often called, may be best illustrated by an example. For this
purpose we will use Fig. 60, which shows a simple truss supporting roof and ceiling, and Fig. 61, which represents a plan of the
walls and trusses. In this case the rafters are supported by purlins,
which come at the joints, while the ceiling beams rest directly on
the tie beam of the truss.

Now if we let D represent the distance in feet from the center
of the space on the left to the center of the space on the right of
any particular truss, then the load at joint 2 for that truss will be
$a \times D \times$ weight per square foot of roof, and at joint 3, $2b \times D \times$
weight per square foot of roof.

The load at joint 6 or 8 will be $c \times D \times$ weight of ceiling per
square foot, and at joint 7, $d \times D \times$ weight per square foot.

The points $x\ x$ and $y\ y$, in Fig. 60, which fix the length of a
and c, should be located half way between the bearings. Thus
the first x is half way between the wall plate and the first purlin,
and the second x is half way between the purlins. The first y
is half way between the wall and the rod at 6, and the second y
half way between the rods at 6 and 7 or 7 and 8.

Example I.—To work out an example in figures, we will assume
that Fig. 60 is a drawing of truss 2 of Fig. 61; then D will equal
13. By measurement we find that a (Fig. 60) equals 11 feet 3
inches, b equals 6 feet 2 inches, c equals 8 feet 3 inches, and d
equals 8 feet 6 inches.

The span of the truss is 33 feet, and the rafters have a rise of
10 inches in 12. The roof is to be of slate on ⅛-inch sheeting;
rafters and ceiling joists 2 x 8 inches, pine or spruce, 16 inches
on centers.

The weight per square foot of roof will therefore be:

	Pounds per Square Foot.
For slate	6½
For sheathing	3
For rafters	3
For purlins	2
For truss	3½
For wind and snow	30
Total	48

The weight per square foot of ceiling will be:

	Pounds per Square Foot.
2 x 8 joist..	3
Lath and plaster.......................................	10
Add	5
Total	18

The load at joint 2 will = $11\frac{1}{4} \times 13 \times 47\frac{3}{4}$ = 6,983 pounds.
Load at joint 3 = $12\frac{1}{8} \times 13 \times 47\frac{3}{4}$ = 7,657 pounds.
Load at joint 6 = $8\frac{1}{4} \times 13 \times 18$ = 1,930 pounds.
Load at joint 7 = $8\frac{1}{2} \times 13 \times 18$ = 1,989 pounds.

The loads at 6, 7 and 8 will be the same, whether the ceiling joists rest on the tie beam, as in Fig. 60, or on purlins hung from the tie beam, as in Fig. 62; but it will make a difference in the required dimensions of the tie beam, as will be explained in Chapter VI.

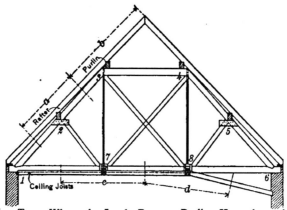

Fig. 62.—Truss Where the Loads Rest on Purlins Hung from the Tie Beam.

When the roof is quite steep it is sometimes desirable to support the roof, as in Fig. 62, without a purlin under the ridge. In this case the load at joint 3 will equal $b \times D \times$ weight per square foot. As a rule, it is better construction to have a purlin at the top.

When the roof truss supports hips or valleys the roof area supported at the joints will vary somewhat, and must be computed from the framing plan of the roof.

TRUSS DIAGRAM FOR TRUSS 60.

Example II.—Having determined the loads at the joints, the next thing to do is to make a skeleton drawing of the truss, which we will call the "truss diagram." This should be made by drawing center lines through the principal members of the truss—in this case the rafters, tie beam and center tie rod. Next draw lines through the center of the rods 2-6 and 4-8. This will give the joints 2 and 4 in Fig. 63, which represents the truss diagram for the truss in Fig. 60. Then draw lines from 2 and 4 to point 7, to represent the braces. These last lines will not quite coincide with the actual center lines of the braces, but in order to draw the stress diagram correctly it is necessary that all the lines in the truss diagram meet at points representing the centers of the joints. In constructing the truss the short ties and braces should be located so that their center lines will intersect on the center lines of the principals as nearly as possible and get a practical joint. When all the lines of the truss diagram are drawn the loads at the various joints should be indicated by arrows and figures, as shown in Fig. 63.* The upper figures at joints 2, 3 and 4, represent the roof loads at these joints, and the figures in the second line the ceiling loads directly below. As far as the stresses in the struts and tie beam are concerned it makes no difference whether the ceiling loads are considered as applied at the top or bottom of the truss, but as it simplifies the stress diagram considerably to consider the ceiling loads as applied at the top, we shall do so in all of the problems explained in these chapters. The roof and ceiling loads should always be put down separately, however, and added together, as in Fig. 63. When the ceiling loads are added to the roof loads they must also be added to the stresses in the vertical rods, as determined by the stress diagram,

---

*In practice it is not worth while to figure the loads closer than even 50 or 100 lbs., as the lines of the stress diagram cannot be scaled much closer than 50 lbs., nor is it practical to proportion the size of the members with a greater degree of accuracy.

because the rods have to transmit this load or stress to the top joint.

If the rods are not plumb the ceiling loads cannot be added to the roof loads, but must be shown separately in the stress diagram. As a reminder to add the ceiling loads to the stresses in the rods, they should be put beside the corresponding rod on the truss diagram, as shown.

MEMBERS NOT SUBJECT TO STRESS.

It often happens that a truss contains members which receive no strain when the truss is considered as loaded only at the top, or when the loads are perfectly symmetrical. Such members cannot be represented in the stress diagram, and hence should be shown in the truss diagram by dotted lines, or if there are no ceiling loads they may be omitted entirely.

The rods 2-6 and 4-8 in the truss, Fig. 60, would not be needed if the loads were all at the top, and should be represented by dotted lines in Fig. 63, as shown. These rods, however, have to support the ceiling loads at 6 and 8, and hence the loads should be indicated beside the dotted lines, as shown. This is the only stress in these rods.

After the loads have been indicated on the truss diagram *the reactions at the ends of the truss* should be computed. When a truss or beam rests on two walls or posts, the latter must afford a resistance upward equal to the weight of the truss and its load. These reactions are considered as forces, and will hereinafter be called the "supporting forces."

For trusses symmetrically loaded—the only kind considered in this chapter—each supporting force will equal one-half of the total joint loads. In the example we are now considering the supporting forces are 13,736 pounds each. The supporting forces should be represented by lines with arrow heads pointing up. They are also often designated by the letters P and P_1.

LETTERING THE TRUSS DIAGRAM.

Finally, the truss diagram should be lettered according to a peculiar method, known as "Bow's notation." The essential principle of this method of lettering is to letter the *spaces* inclosed by the truss members and also the spaces between the external forces so that a line or force on the truss diagram is designated by the letters on each side of it. Fig. 63 is lettered in this way. The bottom half of the rafter on the left is designated as A E. The left-hand brace is E F and the king rod is F G. The tie beam consists of two parts, O E and O H. The supporting force at 1 is O A and the load at 2 is A B.

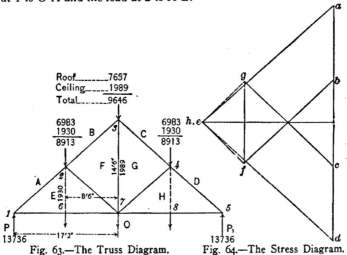

Fig. 63.—The Truss Diagram.   Fig. 64.—The Stress Diagram.

The advantage of this method of lettering is that in the stress diagram the same letters come at the end of the corresponding lines, so that each member and the line representing the stress in it are designated by the same letters. To distinguish the member from its stress capital letters are used on the truss diagram and small letters on the stress diagram. In lettering the truss diagram any letters may be used, in any order, and no attention is paid to members represented by dotted lines.

In actual work it is not necessary to number the joints, as in Fig. 63, but in describing the methods of procedure it is necessary to have some means of indicating the different joints.

DRAWING THE STRESS DIAGRAM.

When the truss diagram is completed, as above described, we can proceed to draw the stress diagram, by which the stresses in the different members of the truss may be determined. The stress diagram is based upon the principle of mechanics, that for a point to be in equilibrium, lines drawn parallel to the forces acting on that point and representing their magnitude to a scale must form a closed polygon, the lines being drawn in the same order as the forces.

Each joint of a truss is supposed to be a pin, and each full line on the truss diagram a force; the forces are supposed to act on the pin, as if they were hinged, and consequently for the joint to stay in its proper place, or to be in equilibrium, the forces acting on any joint must balance each other.

To draw the stress diagram it is necessary to first draw the forces acting on one of the end joints, then those at the first joint above, and thus proceed from joint to joint until all of the forces have been represented. The whole process, when once understood, is quite simple, but is not so easy to explain in a printed article, where the stress diagram must be shown completely drawn. To understand the following description the reader should reproduce on a sheet of paper the truss diagram, Fig. 63, at about twice the size of the illustration, and then draw the stress diagram, line by line, according to the following directions, lettering each line as it is drawn.

The paper should be tacked to a drawing board and all lines in the stress diagram drawn *exactly parallel* to the corresponding lines in the truss diagram.

DRAWING STRESS DIAGRAM FOR FIG. 63.

First draw a vertical line, equal to 13,736 pounds, measured to a scale of, say 4000 pounds to an inch,* and letter the bottom of

---

*For this purpose an engineer's scale, graduated to fortieths of an inch, will be found most convenient.

the line *o* and the top *a*, as shown in Fig. 64. This line will represent the supporting force at joint 1. Next from the upper end *a* draw a line parallel to the rafter A E, and from the bottom end *o* a horizontal line intersecting the other line, and mark the point of intersection *e*. We then have a triangle, *a e o*, which represents the three forces acting on the joint 1. Next, from *a* measure down on *a o*, 8913 pounds, the load at 2, using the same scale of 4000 pounds to the inch, and letter the point thus found *b*. Then from *b* draw a line parallel to B F,* and from *e* a line parallel to

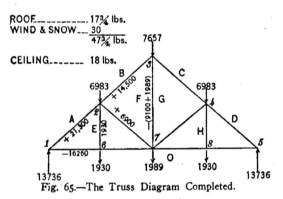

ROOF_____ 17¾ lbs.
WIND & SNOW___ 30
                        47¾ lbs.   7657

CEILING_____ 18 lbs.

Fig. 65.—The Truss Diagram Completed.

E F, lettering the point where the two lines intersect, *f*. The lines *e a, a b, b f* and *f e* then represent the four forces acting at joint 2 (the load being considered as a force) and the number of pounds each force must exert to produce equilibrium.

Next, from *b* measure downward, 9694 pounds, the load at 3, and letter the point thus obtained as *c*. From *c* draw a line parallel to C G, and from *f* a line parallel to F G (vertical), and letter the point where the two lines intersect, as *g*. The lines *f b, b c, c g* and *g f* then represent the forces acting on joint 3. We have now the stress in the king rod, and in all the pieces to the left, and as the truss is symmetrical the stresses in the right side of the truss will be the same as those in the left side, and it is not

*It should be remembered that capital letters refer to truss diagram, Fig. 63, and small letters to the stress diagram, Fig. 64.

necessary to represent them in the stress diagram. If we did so we would get the dotted lines shown in Fig. 64, which would make the diagram symmetrical.

To obtain the stress in the different members of the truss in pounds we measure each line of the stress diagram by our scale of 4000 pounds to the inch, or the same scale that was used in laying off the line *o a*. Doing this, we find *a e* measures 21,300 pounds, *o e*, 16,260 pounds, *f g*, 9100 pounds, and so on.

These figures should now be put on the corresponding lines of the truss diagram, as in Fig. 65. The truss diagram, as finally figured, gives all of the data necessary for determining the size of the pieces. The weights per square foot of roof and ceiling used in computing the joint loads should also be put on the truss diagram for future reference: The stress diagram is now no longer needed and can be destroyed.

The total stress on the center rod is 9100 pounds, obtained by scaling the line *f g*, of Fig. 64, plus 1989 pounds, the load at the bottom, or 11,089 pounds. The sign + before the stresses denotes that the piece is in compression and the minus sign that the piece is in tension.

Example III.—As a further illustration of the method of drawing stress diagrams we will determine the stresses in the truss illustrated by Fig. 66. From the plan and section of the roof we obtain the following data : Span of truss, 36 feet ; distance between trusses, which are uniformly spaced, 15 feet ; distance *a* = 9 feet 10 inches, *b* = 13 feet 4 inches, and *c* = 12 feet 2 inches.

Lines drawn through the center of the truss rafters, the top chord and tie beam give the diagram shown in Fig. 67, the center lines for the braces being drawn to connect joints 2 and 7 and 5 and 8, although they would not exactly correspond with the actual center lines of the braces, for reasons already pointed out. The diagonal braces B B in the center panel are omitted from this diagram because they have no stress when the loads are symmetrical. The roof will consist of 2 x 6 inch spruce rafters, 16 inches on centers, covered with ⅝-inch sheathing and common cedar shingles. The ceiling will be framed with 2 x 6 inch joists,

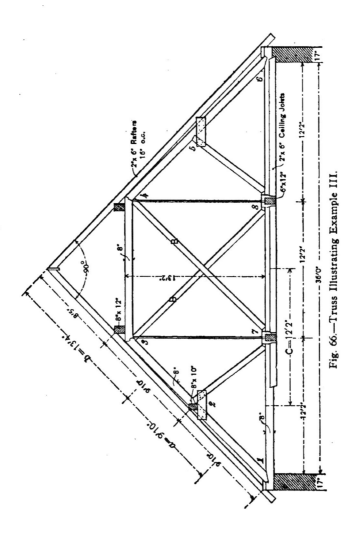

Fig. 66.—Truss Illustrating Example III.

16 inches on centers, lathed and plastered, and supported by purlins at joints 7 and 8. The weight per square foot of roof will be
as follows:

Shingles	2½ pounds
Sheathing	3 pounds
Rafters	2¼ pounds
Purlins	2 pounds
Truss	3 pounds
Allowance for wind and snow....	30 pounds

Total .... 42¾ pounds per square foot
Allowance for ceiling............ 20 pounds per square foot
Using the above figures, the roof load at joint 2
  will be 9 5-6 × 15 feet × 42¾ pounds, or....... 6,306 pounds
Load at 3, 13 1-3 × 15 feet × 42¾ pounds, or..... 8,550 pounds
Load at 7, 12 1-6 × 15 feet × 20 pounds, or....... 3,650 pounds

These loads should then be placed on the truss diagram, as
shown, and the loads at 7 and 8 put beside the vertical rods, to
show that they are to be added to the stress given by the stress
diagram, as explained under Example II.

Each supporting force will equal one-half of the load on the
truss, or 18,506 pounds. The diagram should be lettered according to the principle previously explained and as shown on the
diagram.

TO DRAW THE STRESS DIAGRAM.

On the same sheet of paper as the truss diagram, and a little to
the right, draw a vertical line, *o a*, equal in length to the supporting force P (18,506 pounds), at a scale of, say, 4000 pounds to the
inch.* The bottom of this line should be lettered *o* and the top *a*.
Then from *a* draw a line parallel to the rafter A F and from *o*
a horizontal line intersecting the first line, and mark the point of
intersection *f*.†

*To thoroughly understand the method of drawing the stress diagram,
the reader should take a sheet of paper on a drawing board, and with
T-square, scale and triangle draw the truss diagram, for which all necessary measurements are given in Fig. 67, with the utmost accuracy. Having
drawn the truss diagram, draw the stress diagram line by line, according
to the directions, being careful to draw the lines perfectly parallel with
those in the truss diagram.

†The reader will please remember that small letters refer to lines in
the stress diagram and capital letters to lines in the truss diagram.

The reason the point of intersection is lettered f is because F is the other letter designating the lower half of the rafter and the left-hand end of the tie beam.

The lines a f and f o show the stress in the corresponding members of the truss.

We must next consider the forces acting at joint 2, of which there are four. One of these, f a, is already drawn. For A B we measure downward from a a distance equal to 6306 pounds and letter the point thus obtained b. To find the stresses in B H and H F we draw from b an indefinite line parallel to B H, and from f a line parallel to F H, until it intersects the line from b, and letter the point of intersection h.

The next step is to draw the stresses at joint 7. Of the stresses acting at this joint we already have o f and f h, and for the other two we draw a vertical line through h, and at the point where it intersects the line o f place the letter k. The polygon of forces for joint 7 is then o f, f h, h k and k o. Although k o lays over f o, it should be considered as a separate line.

Next determine the polygon of forces for joint 3. We already have k h and h b, so from b measure downward 12,200 pounds, which in this case comes at o; but we also put here the letter c, as b c represents the load at 3.

There now remains only one unknown force, C K, and as we already have the point k, the line c k must represent this force or stress, thus showing that the stress in C K is the same as in O K. As the truss is symmetrically loaded, the stresses in the other half of the truss must be the same as in the half we have determined, hence it is not necessary to carry the stress diagram further.

Applying our scale of 4000 pounds to the inch to the lines of the stress diagram, we obtain the values given in Fig. 68, which should either be put on the stress diagram or on the corresponding members of the truss diagram. To the stress obtained by measuring the line h k must be added the ceiling load at 7, thus making the actual stress in the rods 6930 pounds. How to proportion the timbers and rods to the stresses will be explained in Chapter VI.

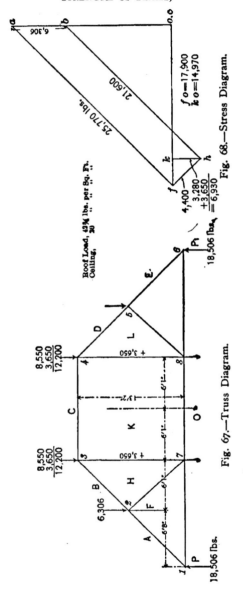

Fig. 68.—Stress Diagram.

Fig. 67.—Truss Diagram.

Example IV.—We will now determine the stresses in the horizontal truss, Fig. 69. The trusses are supposed to be uniformly spaced, 15 feet 8 inches center to center, and the rafters and ceiling joists span from truss to truss and rest directly on the top and bottom chords, as the horizontal pieces are called. The weight per square foot of roof surface we find to be as follows (see Tables IX to XII), the building being situated in the Central States:

	Pounds.
Five-ply gravel roofing	6
Sheathing	3
Rafters	3
Truss	4½
Snow	30
Total	46½

The roof being flat, there will be no wind pressure. The load at joint 2 will equal

$$\frac{8 \text{ feet } 4 \text{ inches} + 7 \text{ feet } 10 \text{ inches}}{2} \times 15 \text{ feet } 8 \text{ inches} \times 46\tfrac{1}{2} \text{ pounds}$$

$= 5885$ pounds.

Load at joint 4 equals 7 feet 10 inches × 15 feet 8 inches × 46½ $= 5704$ pounds.

Load at joint 6 equals that at 4, or 5704 pounds.

The actual weight of the ceiling joists, lath and plaster will be about 13 pounds, but we will allow 3 pounds extra and use 16 pounds per square foot for weight of ceiling.

The load at 3 will then equal

$$\frac{8 \text{ feet} + 7 \text{ feet } 10 \text{ inches}}{2} \times 15 \text{ feet } 8 \text{ inches} \times 16 = 1984 \text{ pounds.}$$

Loads at 5 and 7 equal 7 feet 10 inches × 15 feet 8 inches × 16 $= 1968$ pounds.

The truss diagram will be as shown in Fig. 70, and the joint loads should be indicated as shown, the ceiling loads being added to the roof loads, as explained under Example II. The figures beside the truss lines in Fig. 70 are the stresses, obtained from Fig. 71, and, of course, cannot be put on the truss diagram until the stress diagram is completed.

The supporting forces will each be equal to one-half the sum of the joint loads, or 19,377 pounds.

In lettering the truss diagram it should be borne in mind that the center rod supports only the ceiling load at 7, and hence cannot appear in the stress diagram when the loads are considered as applied at the top. The line representing the center rod should therefore be dotted, and the same letter should be used on both sides of the dotted line.

To construct the stress diagram, we begin by drawing the supporting force at joint 1, which is represented by the vertical line *o a*, Fig. 71. From *a* draw an indefinite line parallel to A H, and through *o* a horizontal line until it intersects the first line. This

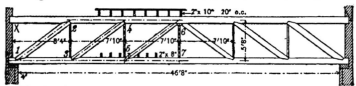

Fig. 69.—Horizontal Truss.—Spacing of Trusses 15 ft. 8 ins. C to C.

point of intersection should be lettered *h*, because H is the letter common to the brace and tie beam in the end panel. The triangle *o a h o* then represents the forces or stresses acting at joint 1.

Next draw the polygon of forces at joint 2. We already have the stress in A H, represented by *h a*, and from *a* we measure downward to *b,* a distance equal to 7869 pounds—the load at 2. Then through *b* draw a line parallel to B K, and through *h* a line parallel to H K, until they intersect, and letter the point of intersection *k*. The figure *h a b k h* is then the polygon of forces for joint 2.

We must next complete the polygon for joint 3. Here we already have the lines *o h* and *h k*, and from *k* we draw a line parallel to K L and extend the line *o h* until the two intersect, lettering the point of intersection *l*. The lines *o h, h k, k l* and *l o* then represent the forces acting at joint 3.

Next draw the polygon for joint 4, which is formed by the lines *l k, k b, b c, c m* and *l m.*

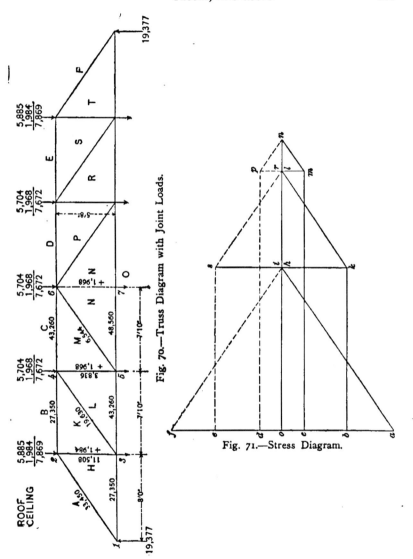

Fig. 70.—Truss Diagram with Joint Loads.

Fig. 71.—Stress Diagram.

For the polygon forces at joint 5 we have *o l* and *l m*, and we draw *m n* and *o n*. We now have the stress diagram for all the truss members to the left of the center, and as the stresses in the members on the right side must necessarily be of the same magnitude, it is not necessary to go further with the stress diagram. By continuing the process, however, we would obtain the dotted lines below *o n*, making the complete stress diagram, which if correctly drawn will be symmetrical about the line *o n*.

Measuring the lines in the stress diagram by our scale, we obtain the stresses given on the corresponding lines of the truss diagram. The first of the figures on the vertical lines are the stresses obtained from the stress diagram, and the other number is the load at the joint below, which must be added to the stress to give the full strain in the rod.

COMPARISON OF THE GRAPHIC METHOD WITH THE RULES GIVEN ON PAGES 64 TO 68.

The stresses in the truss shown in Figs. 69 to 71 inclusive can be computed by Rules 1 to 10, given on page 64, and it will be good practice for the student to make the computations and compare them with the stresses obtained by the graphic method. Fig. 72 gives the necessary data for making the computation. By the

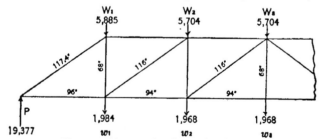

Fig. 72.—Diagram for Computing Stresses.

graphic method it is impossible to determine the stresses below 100 pounds with much accuracy, as with a scale of 4000 pounds to the inch 100 pounds is a very small division, and if a scale of 1000 pounds to the inch is used it makes the lines so long that it is difficult to get them absolutely parallel to the lines of the truss diagram. As a rule the author uses such a scale of pounds as will make the line representing one supporting force from 3½ to 6 inches long. For roof trusses a variation of 100 pounds in the stress of any member would not affect the proportioning of the parts.

SIMPLE SCISSORS TRUSS.

Example V.—We will next determine the stresses in a simple scissors truss, as shown in Fig. 73. This truss is intended to

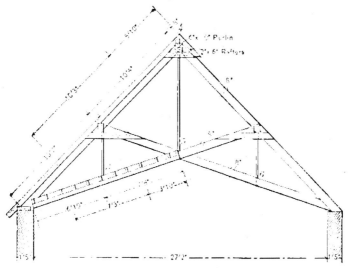

Fig. 73.—Simple Scissors Truss Where the Trusses are Placed 12 Feet on Centers.

support a shingle roof and plastered ceiling, and the weight per square foot of roof surface will be as follows:

For Shingles .... 2½ pounds
Sheathing .... 3 pounds
Purlins .... 2 pounds
Rafters .... 2¼ pounds
Truss .... 3 pounds
Wind or snow..... 30 pounds
Total .... 42¾, or say 43 pounds per square foot

The actual weight of the ceiling would be about $12\frac{1}{4}$ pounds per square foot, but to allow for possible additional loading we will allow 16 pounds per square foot. The trusses to be spaced 12 feet apart from centers. The roof load at joint 2 will equal

$$10 \text{ ft. } 3 \text{ in.} \times 12 \text{ ft.} \times 43 \text{ lbs.} = 5289 \text{ lbs.}$$

and at joint 4,

$$11 \text{ ft. } 8 \text{ in.} \times 12 \text{ ft.} \times 43 \text{ lbs.} = 6020 \text{ lbs.}$$

The ceiling load at 3 will equal

$$7 \text{ ft. } 3 \text{ in.} \times 12 \text{ ft.} \times 16 \text{ lbs.} = 1382 \text{ lbs.}$$

and at joint 5,

$$7 \text{ ft. } 8 \text{ in.} \times 12 \text{ ft.} \times 16 \text{ lbs.} = 1472 \text{ lbs.}$$

As the ceiling loads are supported by vertical rods, we can assume that they are applied at the upper joints as in the preceding examples.

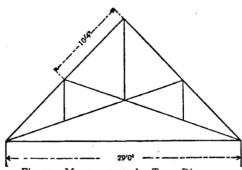

Fig. 74.—Measurements for Truss Diagram.

The necessary dimensions for drawing the truss diagram representing the center lines of the truss members are given on the diagram, Fig. 74, the inclination of the rafters being $45°$.

In Fig. 75 is represented the truss diagram properly lettered and with the loads indicated. Each supporting force is equal to one-half the sum of the joint loads.

In order to construct the stress diagram we first draw a vertical line, $o\ a$, Fig. 76, equal to and representing the supporting force at joint 1. A scale of 3000 pounds to the inch will be the most suitable for this diagram. Next, through a draw an indefinite

line parallel to A D of Fig. 75 and through *o* a line parallel to O D
and intersecting the first line. Letter the point of intersection *d*.
Then *a d* represents the stress in the bottom of the rafter, and *o d*
the stress in the tie beam.

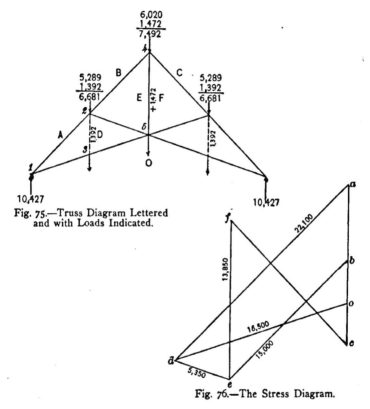

Fig. 75.—Truss Diagram Lettered
and with Loads Indicated.

Fig. 76.—The Stress Diagram.

We must next complete the polygon of forces for joint 2. The
first line of this polygon will be the line *d a,* already drawn. From
a measure downward 6681 pounds, the load at 2, and letter the
point thus obtained *b*. Through *b* draw a line parallel to B E,
and through *d* a line parallel to D E, and letter the point of inter-

section *e*. The four-sided figure *d a b e d* is the polygon of forces for joint 2. The vertical rod 2-3 merely transmits the ceiling load to the rafter, and cannot be represented in the stress diagram, or in other words, the stress diagram is drawn as though there were no such rod in the truss.

We must next draw the polygon of forces for joint 4. The first line of this polygon, *e b*, is already drawn. From *b* measure downward 7492 pounds, the load at 4, and letter the point thus obtained *c*. Through *c* draw a line parallel to C F, and through *e* draw a line parallel to E F, and letter the point of intersection *f*.

We now have the stress in the center rod, and for all the members to the left. The stresses in the truss members to the right will be the duplicate of those already found. If the stress diagram has been correctly drawn, a horizontal line through *o* will bisect the vertical line *e f*. Measuring the lines of the stress diagram we obtain the figures given on the lines of the diagram, which are the stresses in the corresponding truss members in pounds. The total strain in the center rod will be 13,850 pounds plus the load at 5 (1472), or 15,332 pounds. The stress in each of the other two vertical rods will be merely the ceiling load at 3 or 6, or 1392 pounds.

This method of determining the stresses may be applied to all symmetrical trusses in which the ceiling loads are transmitted to the top of the truss by vertical rods. When the rods are not vertical, but inclined, the ceiling loads cannot be added to the roof loads, and it is necessary to show the ceiling loads separately in the stress diagram. The stresses in any true truss may be determined by means of stress diagrams, but as the trusses become elaborate the method of drawing the tress diagram becomes more complicated, and should not be undertaken without a thorough understanding of the subject.

Stress diagrams for almost all forms of trusses are given in "The Architects' and Builders' Pocket Book," and the entire subject of roof truss design is very fully treated in Part III of "Building Construction and Superintendence."

CHAPTER VI.

HOW TO COMPUTE THE SIZES OF TRUSS MEMBERS.

The safe and proper dimensions of any member* of a truss cannot be fixed with any degree of accuracy until the stress which the maximum load on the truss will produce in that member is known. The method of determining the stresses in simple trusses, symmetrically loaded, has been explained in Chapter V, and we will now endeavor to show how to proportion the members to the stresses.

In a well designed truss every member will be either in tension or in compression, and some of the members may also be subjected to a transverse or cross strain. This gives us four kinds of members—namely:

TIES, which are in tension only.

STRUTS, which are in compression only.

TIE BEAMS, which are subject both to tension and cross strain.

STRUT BEAMS, subject to compression and cross strain.

In computing the size of any one of the above the first step should be to determine the least area or cross section required to resist the stress, and then increase the size of the piece sufficiently to allow for bolt holes and for making practical connections at the joints.

The following rules and tables will enable the reader to determine the least net sectional area of any member:

TIES.

Rule 1.—To find the net or least sectional area of a tie divide the stress in pounds by the following values, and the answer will be in square inches:

---

*The pieces of wood, iron or steel composing a truss, and extending from one joint to another, are called "members." Bolts or straps used in making joints and superfluous pieces are not included in this term.

White pine	1,400
Spruce	1,600
Norway pine	1,600
Oregon pine	1,800
Long leaf yellow pine	2,000
Wrought iron	12,500
Steel	15,000

The strength of a tie is not affected (theoretically) by its length nor by the shape of the cross section.

Thus, if the net area required in a wooden tie is 12 square inches the tie may be made either 3 x 4 inches, or 2 x 6 inches, or 1 x 12 inches. The 2 x 6 or 1 x 12 shape, however, will be better for making the connection at the joints. If the material composing the tie were perfectly homogeneous from end to end the strength would not be affected by the length, and in computing the size of wrought iron or steel ties the length is not usually taken into account. In a timber tie a long piece of timber would generally contain more knots than a shorter piece, and in this way its strength be reduced. When rods are used for ties the size of the rod may be determined directly from the following table:

TABLE XIII.—*Safe Loads for Round Rods of Wrought Iron and Steel.*

(Based on 12,500 pounds per square inch for iron and 15,000 pounds for steel.)

Diameter of rods. in inches.	Rods not upset.		Rods with raised threads or upset.	
	Wrought iron.	Steel.	Wrought iron.	Steel.
½	1,570	1,884	2,453	2,944
⅝	2,453	2,944	3,835	4,600
¾	3,750	4,500	5,520	6,627
⅞	5.250	6,300	7,516	9,020
1	6,780	8,140	9,815	11,780
1⅛	8,570	10,290	12,425	14,900
1¼	11,060	13,270	15,330	18,400
1⅜	13,370	16,050	18,550	22,260
1½	16,080	19,300	22,080	26,500
1⅝	18,750	22,500	25,910	31,090
1¾	23,000	26,400	30,060	36,070
1⅞	25.250	30,300	34,500	41,400
2	28,500	34,200	39,270	47,130
2⅛	33,000	39,600	44,320	53,190
2¼	37,500	45,000	49,700	59,680
2⅜	41,870	50,250	55,370	66,450
2½	46,500	55,800	61,350	73,620

STRUTS.

The resistance of a strut, unlike that of a tie, is affected by its length, and consequently the method of determining the safe strength of a strut is much more complicated. By means of the following table, however, one can find the necessary size of strut to use for almost any stress or length that would occur in ordinary wooden trusses. No additional resistance should be allowed for lengths less than those for which values are given in the table.

TABLE XIV.—Maximum Safe Loads for Wooden Struts in Pounds.

Struts of white pine or spruce.

Size in inches.	6	8	10	12	14
2 x 6	4,900	3,040	....	....	....
2 x 8	6,540	5,390	....	....	....
3 x 6	8,650	7,790	6,930	....	....
3 x 8	11,540	10,390	9,240	....	....
4 x 4	8,270	7,690	7,120	6,540	....
4 x 6	12,400	11,520	10,550	9,800	8,700
4 x 8	16,540	15,390	14,240	13,080	11,930
4 x 10	20,640	19,240	17,800	16,360	14,920
6 x 6	19,900	19,080	18,216	17,352	16,490

Size in inches.	8	10	12	14	16
6 x 8	25,440	24,290	23,140	21,980	20,830
6 x 10	31,800	30,360	28,920	27,480	26,040
8 x 8	35,450	34,300	33,150	32,000	30,850
8 x 10	44,320	42,480	41,440	40,000	38,560
8 x 12	58,180	51,450	49,730	48,000	46,270
10 x 10	62,500	55,400	53,960	52,520	51,080
10 x 12	75,000	66,480	64,800	63,000	61,300
12 x 12	....	90,000	79,780	78,000	76,320
12 x 14	....	105,000	93,170	91,050	89,000

Struts of Oregon pine (Douglas fir) and long leaf yellow pine.

Size in inches.	6	8	10	12	14
2 x 6	7,680	6,240	....	....	....
2 x 8	10,240	8,320	....	....	....
3 x 6	13,680	12,240	10,800	9,360	....
3 x 8	18,240	16,320	14,400	12,480	....
4 x 4	13,120	12,160	11,200	10,240	9,280
4 x 6	19,680	18,200	16,800	15,360	....
4 x 8	26,240	24,300	22,400	20,480	18,560
4 x 10	32,800	30,400	28,000	25,600	23,200
6 x 6	31,680	30,200	28,800	27,400	25,900

Size in inches.	Length in feet.				
	8	10	12	14	16
6 x 8	40,300	38,400	36,500	34,600	32,600
6 x 10	50,400	48,000	45,600	43,230	40,800
8 x 8	64,000	54,400	52,500	50,600	48,600
8 x 10	80,000	68,000	65,600	63,200	60,800
8 x 12	96,000	81,600	78,700	76,800	73,000
10 x 10	....	100,000	85,600	83,200	80,800
10 x 12	....	120,000	102,700	99,800	97,000
12 x 12	....	....	144,000	123,800	121,000
12 x 14	....	....	168,000	144,500	142,800

TIE BEAMS AND STRUT BEAMS.

Determining the size of a tie beam or strut beam involves two computations—first, finding the size of beam to resist the cross strain, and, second, the sectional area required to resist the tension or compression, and then adding the two together.

TO FIND THE SIZE OF BEAM TO RESIST THE CROSS STRAIN.

Rule 2.—When the load is distributed over the whole length of the beam: Assume the depth, multiply the length, or span, in feet by the load in pounds, and divide by twice the square of the depth multiplied by the value for A given in Table I, page 5. The answer will be the breadth of the beam in inches.

If the load is concentrated at the center of the span multiply by 2 and proceed as above.

If the load is concentrated at one-third of the span multiply by 1.78, if at one-fourth of the span multiply by 1.5, and if at one-fifth of the span, by 1.28. (See page 11, for factors giving equivalent distributed load.)

Example I.—The application of the above rules to tie beams will be more clearly shown in connection with the following example, where we will determine the size of the members in the truss shown by Fig. 60 and for which the stresses are given in Fig. 65.

RODS.

The stress on the center rod is 9100 + 1989, or 11,089 pounds. Assuming that the rods will be of wrought iron with the screw

thread cut from the body of the rod, we find from Table XIII
that a 1¼-inch rod has a safe strength of 11,060 pounds, or 29
pounds less than the stress. As a large allowance was made for
wind and snow in the roof loads we may safely use the 1¼-inch
rod. The short rods E and H, Fig. 65, have a stress of 1930
pounds each, which will require a ⅝-inch rod.

STRUTS.

The lower portion of the main strut or rafter has a stress of
21,300 pounds, and is about 10 feet long between joints 1 and
2. Supposing the wood to be white pine we find from Table XIV
that a 6 x 8 timber 10 feet long has a safe resistance of 24,290
pounds, hence this size will answer for the rafter. The stress in
the upper half of the rafter is not as great as in the lower portion,
but on account of the construction of the joints it will be much
better and about as cheap to make the rafters 6 x 8 for their full
length, rather than to use a smaller section for the upper portion.

The braces E F and G H have a compressive stress of 6900
pounds each, and are of the same length, 10 feet.

From Table XIV we find that a 4 x 4, 10 feet long, has a safe
resistance to crushing of 7120 pounds, but as it will be better to
have the brace of the same width as the rafter we will make the
braces 4 x 6 inches.

TIE BEAMS.

We now come to the tie beam. The direct tension in the tie
beam is 16,260 pounds. By Rule 1 we find that the sectional area
required to resist this tension is 16,260 ÷ 1400, or 11.6 square
inches, or say, 12 square inches. This is equivalent to 1½ x 8
inches, or 2 x 6 inches.

The tie beam also supports the ceiling joists, and we must compute the sectional area required to resist the transverse strain.

As the tie beam is supported by the rods the span of the beam
should be taken as the distance between two adjacent joints. In

this case the longest span is from 6 to 7, or 7 to 8. This distance is 8 feet 6 inches. The load on this span will be the same as that at joint 7, or 1989 pounds. We determine the size of beam required to support this load by Rule 2. Assume the depth as 8 inches. The span multiplied by the load $(8\frac{1}{2} \times 1989) = 16,907$. Twice the square of the depth multiplied by the value of A for white pine, Table I $(2 \times 64 \times 60) = 7680$. 16,907 divided by $7680 = 2\frac{1}{4}$ inches, nearly. Hence, to resist the cross strain will require a beam $2\frac{1}{4} \times 8$ inches. As we found that it will require $1\frac{1}{2} \times 8$ inches to resist the direct tension the beam must have a sectional area equal to their sum, or $3\frac{3}{4} \times 8$ inches. It is desirable, however, that the tie beam shall be as thick as the rafter, on account of making a good joint at the support. We will therefore see if a 6 x 6 will answer.

Fig. 77.—Plan of 6 x 8 Tie Beam, Showing Manner of Breaking Joints.

Assuming the depth as 6 inches we would have for our divisor, under Rule 2, $2 \times 36 \times 60 = 4320$. This is contained in 16,907 3.9 times, or practically 4 inches. Our area of 12 square inches to resist tension is equal to 2 x 6 inches, hence a 6 x 6 inch timber will just answer for the tie beam. If we use this dimension, however, the tie beam cannot be spliced, but must be in one piece, 34 feet 4 inches long. If a timber of this length cannot be readily obtained we must increase the size to 6 x 8 inches, and build the beam of three thicknesses of 2 x 8 inch planks, breaking joint as shown in Fig. 77, and spiked and bolted together. This gives a net area between the rods of 4 x 8 inches, which we found was sufficient, and opposite the center rod of $2\frac{5}{8} \times 8$ inches. As there is no cross strain at the joints we there require only sufficient area to resist the direct tension.

When tie beams are built up it is very important to arrange the joints of the planks so that the net area will be everywhere sufficient to resist the strain and to give space for enough bolts and spikes to transmit the strain from one set of planks to another.

The correct method of splicing built-up tie beams is given in Chapter VII. It is always best to use a single stick for the tie beam whenever practicable.

After the size of all the truss members has been determined they should be put on the figured truss diagram as shown in Fig. 78, which is Fig. 65 of Chapter IV, with the sizes of members added. This gives all of the data for the truss, except the details of the joints, which will be considered in Chapter VII.

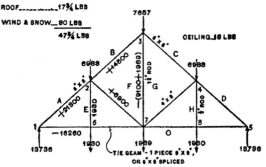

Fig. 78.—Truss Diagram Completely Figured.

Example II.—As a further illustration of the application of the rules we will compute the size of the members of the truss shown in Fig. 66, and for which the stresses are given in Fig. 68.

RODS.

The stress in each of the two rods is 6930 pounds. Assuming that they will be of wrought iron, with threads cut from the body of the rod, we find from Table XIII that a 1-inch rod is not quite strong enough, and that we must use 1⅛-inch rods.

PRINCIPAL STRUTS.

The stress in the lower portion of the principal struts or rafters is 25,770 pounds. We will assume that the timber is to be of white pine. The length between joints 1 and 2 is a little less than

10 feet. From Table XIV we find that to resist a compression of 25,770 pounds, with a length of 10 feet, will require either a 6 x 10 or 8 x 8 inch strut. Which of these sizes we will use we can determine better after we have found the required size for the top chord and tie beam. The stress in the top chord is 14,970 pounds and the length is practically 10 feet. From Table XIV we see that a 6 x 6 inch strut has a safe resistance for this length of 17,352 pounds, so that it will do for the top chord, but as a 6 x 8 inch will make a better joint at 3, and will cost but very little more, we will make the top chord 6 x 8. The two braces, 2-7 and 5-8, each have a compressive stress of 4400 pounds and are about 9 feet long. From Table XIV we see that this will require a 3 x 6 inch timber. The tie beam in this truss has no transverse strain to speak of. The tensile stress is 17,900 pounds. From Rule 1 we find the net sectional area by dividing the stress by 1400—for white pine—which gives 12.8 square inches, or a little more than 2 x 6 inches. As the tie beam will probably be built up of planks we must double the net area to allow for splicing, and it will also be necessary to still further increase the section to allow for cutting at the joints and for the bolts and rods, so that we will make the size of the tie beam 6 x 6 inches. In order to make good connections at the joints the various timbers should all be of the same width, hence we will make the principal struts 6 x 10 inches. The braces B B would have no stress except under wind pressure. What the stress would be in that case can be determined by a special diagram, but as it would not be very large we will be perfectly safe in making them 4 x 6 inches.

Example III.—We will next compute the size of the members for the truss shown in Fig. 79, for which the loads and stresses are given in Fig. 80. [This is the same truss shown by Figs. 69 and 70.]

RODS.

The stress in the rod 2-3 is 13,492 pounds; in rod 4-5, 5804 pounds, and in the center rod 1968 pounds. These will require for wrought iron 1⅜, 1 and ⅝ inch rods respectively. The ⅝-inch rod had better be increased to ¾-inch for appearance sake.

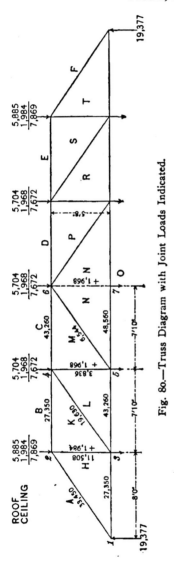

Fig. 80.—Truss Diagram with Joint Loads Indicated.

Fig. 79.—Horizontal Truss.

Of the wooden members we will first compute the size of the top chord, which acts both as a beam and as a strut. The load on one of the center spans is 5704 pounds. Assuming that the wood will be white pine and taking 10 inches as the trial depth, we find the breadth to resist the cross strain by Rule 2. The length or span is about 7¾ feet. Multiplying this by the load we have 44,-206. This is to be divided by twice the square of the depth multiplied by 60 (the value of A for white pine), or 12,000. Performing the division we have 3.7 inches as the breadth, or it will require a beam 3.7 x 10 inches to resist the cross strain. The compression in the center spans of the top chord is 43,260 pounds.

From Table XIV we see that the strength of an 8 x 10 inch beam 8 feet long is about 1000 pounds in excess of this. Therefore to resist the compression will require an 8 x 10 inch beam and to resist the cross strain nearly a 4 x 10 inch, so that our beam must be 12 x 10 inches. As the beam will be stronger, however, if we make the depth 12 inches and the width 10 inches we will use a 10 x 12 inch for the top chord. This might be reduced between the wall and the end joints to 6 x 12 inches. It is not necessary to increase the size of the compression members on account of bolt holes or splices, provided that they come at the joints.

We will next compute the size for the tie beam, which also has a transverse load. The load on one of the center spans of the tie beam is 1968 pounds. Assuming the depth as 10 inches and computing the thickness in the same manner as for the top chord, we obtain 1¼ inches for the necessary thickness of a 10-inch beam to resist the transverse strain. The tension in the center panels is 48,560 pounds.

Dividing by 1400 we have 35 square inches as the net sectional area required to resist the tensile stress. This is equal to 3½ x 10, which is the least sectional area that will answer at joints 5 or 7. Between the joints the thickness must be increased by 1¼ inches to allow for the transverse strain, so that a beam 4¾ x 10 inches is the least that will answer for a solid tie beam—that is, formed of a single stick of timber. As it would be impracticable to obtain such a timber in most localities it will be necessary to build the beam of

2 x 10 inch planks, bolted together every 2 feet with ¾-inch bolts. On account of the joints due to the splicing of the planks it will be necessary to double the size of the beam or to make it 10 x 10 inches, and even then it will be necessary to lay out the beam in such a way that the net sectional area at any point will be at least 35 square inches and so that no two joints in the planks will come nearer than 8 feet of each other. This matter of building up tie beams is considered in the next chapter.

We have now settled on 10 x 12 inches as the size for the top chord and 10 x 10 inches for the tie beam. Brace 1-2 must, therefore, be 10 inches wide. The stress in this brace is 33,450 pounds, and the length of the brace is about 9 feet. From Table XIV we find a 6 x 10 timber is hardly strong enough, and that we must use an 8 x 10, which should be placed flatways. The stress in the brace 3-4 is 19,630 pounds, for which we will have to use a 10 x 6 or a 10 x 5. For brace 5-6 the stress is 6544 pounds, for which a 3 x 6 would answer, but as the beams are 10 inches wide it would be better to use a 3 x 8.

This gives all of the dimensions of the truss, as both sides of the truss should be alike. With these examples the reader should be able to compute the size of the members in any truss after the stresses have been determined, as the process is the same, no matter what the shape of the truss may be.

CHAPTER VII.

PROPORTIONING THE JOINTS OF WOODEN ROOF TRUSSES.

As the strength of a structure is measured by the strength of its weakest part, it is evident that unless the joints of a truss are as strong as the members of which it is composed the truss will not be safe for the load it was intended to carry.

To design the joints of a wooden truss so that they will be capable of transmitting the full stress from one member to another often requires more thought and skill than to compute the size of the members, and the weakest point in most timber trusses is usually at one of the joints.

The correct method of computing the strength of the various joints in wooden trusses can best be shown by considering each kind of joint separately and illustrating by practical examples.

We will first consider the joint formed by the principal strut of a king or queen truss, or of a truss with a horizontal top chord, with the tie beam.

The simplest form of such a joint is that shown in Fig. 81, or in case one bolt is not sufficient then two bolts may be used, as shown in Fig. 82. The strap joint shown in Fig. 83 is also frequently used for light trusses. Fig. 84 shows a better and stronger joint than that shown in Fig. 82, while the strap joint in Fig. 85 is also a stronger joint than that shown in Fig. 83. In Fig. 86 is shown a plate joint, which is suitable for heavy trusses. The joints shown by these six figures may be considered as representative types, although variations of them will often be seen.

In order to understand the principles involved in proportioning any of these joints to the stresses in the rafter and tie beam, one should comprehend the manner in which they are likely to fail. Fortunately a series of tests have been made on joints like those shown, which were carried to the point of destruction,* and from

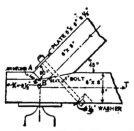

Fig. 81.—Simple Form of Joint.   Fig. 82.—Simple Joint with Two Bolts.

these tests we can gain a pretty good idea of the strength and weakness of such joints. A joint made exactly as in Fig. 81, the timber being of hard pine, failed first by the shearing off of the shoulder on the tie beam and then by the breaking of the bolt.

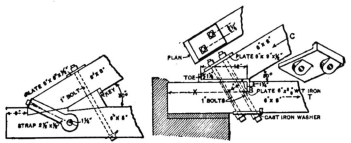

Fig. 83.—Strap Joint.   Fig. 84.—A Stronger Form of Joint than that Shown in Fig. 82.

The stress in the strut at the time of the first failure was 72,300 pounds.

*The tests referred to were made at the Massachusetts Institute of Technology in 1897. A description of the tests and analysis of the results was published by the writer in the *Engineering Record* of November 17, 1900.

The joint shown in Fig. 82 first failed by the shearing of the tie beam, then by the breaking of the inner bolt and finally the outer bolt broke. The joint shown in Fig. 83 broke by shearing of the tie beam and then by the breaking of the strap. A view of the joint after the strap broke is shown in Fig. 87. The end of the strut, where it bore against the shoulder, was considerably splintered, as shown in the illustration. A joint, like that shown in Fig. 84, but without the cast washers on the bottom of the tie beam, failed by the pulling apart of the tie beam opposite the bolts, as shown in Fig. 88. This joint developed a strength 70 per cent. greater than the joint shown in Fig. 82, although the bolts and timbers and the inclination of the strut were the same. The compression in the rafter at the time of failure was 82,900 pounds.

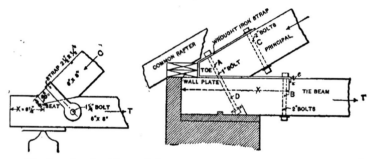

Fig. 85.—Another Form of Strap   Fig. 86.—A Plate Joint for Heavy
            Joint.   Trusses.

The cutting of the washers into the bottom of the tie beam must have materially weakened the beam, so that had a cast washer, as in Fig. 84, been used, still greater strength would probably have been developed. A joint, similar to Fig. 86, but with only one bolt at B and C, failed first by the lug on the plate drawing out of the tie beam, as shown in Fig. 89. For some time before complete failure the bolts B and D (Fig. 86) had been shearing through the tie; finally the bolt B broke by tension and bending. In the joint tested the bolt B was placed some 3 or 4 inches back from the end of the strap. Had it been placed close to the end, as in Fig. 86, the lug could not have pulled out and a much greater strength in the joint would probably have been developed.

In all of the joints where the rafter butted against a shoulder on the tie beam the first failure was by the pushing off, or shearing, of the portion of the tie beam beyond the notch. As soon as this was done the whole stress in the strut was, of course, thrown on the bolt, or bolts, or strap. In some cases the resistance of the shoulder was sufficient to crush or splinter the toe of the strut. In the case of joints 84 and 86 there was no shearing, and the failure was due either to the breaking of the bolts or to the pulling apart of the tie beam. In the joints shown in Figs. 81, 82 and 83 the bolts or strap and the shoulder of the tie beam both united in resisting the thrust of the strut, but just what part was borne by the shoulder and what part by the bolts or strap it is impossible to determine.

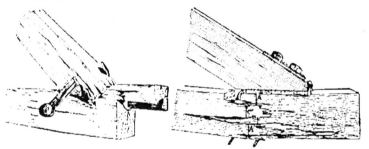

Fig. 87.—Showing Tie Beam Pulled Apart. Fig. 88.—Showing How Joint in Fig. 83 Gave Way.

In designing such joints, therefore, the distance X should either be sufficient to resist the entire thrust or else no dependence whatever should be placed upon the shoulder, and the bolt or strap should be made strong enough to resist the entire stress. There are undoubtedly many trusses standing to-day in which the joints are not made on this principle, but it is more from good fortune than from any precaution on the part of the builder or designer.

The resistance of the shoulder to being pushed off, as in Fig. 87, is known as the resistance to longitudinal shearing, and is due to the adhesion of the fibers of the wood. As the number of fibers are proportional to the area sheared, it is evident that the greater the distance X the greater will be the resistance of the shoulder.

Rule 1.—*To find the distance X necessary to resist the entire thrust of the strut,* divide the stress in the tie beam in pounds by the breadth of the tie beam multiplied by the value for F given in Table XV. The result will be in inches.

Where the shoulder is depended upon to resist the thrust of the strut, it is also necessary to proportion the depth of the notch A B, Fig. 81, so that the toe of the strut or the face of the shoulder will not be splintered by compression.

Fig. 89.—Showing Failure of Joint Similar to Fig. 6.

Table XV.—Values for Constants to be Used in Connection with the Rules for the Strength and Proportion of Truss Joints. All Values Are in Pounds.

	Longitudinal Shear.			Crushing.		
Wood	Wood free. F.	Wood under compression. F_1.	Cross shear. F_2.	End-ways C_1.	Across the grain. C_2.	Trans-verse strain. A.
White oak....	150	250	1,000	1,350	600	75
Yellow pine...	125	250	1,200	1,500	500	100
Oregon pine..	125	250	900	1,350	400	90
Spruce	90	180	750	1,200	300	70
White pine...	80	160	500	1,000	250	60

Rule 2.—*To prevent crushing the toe of the rafter* the depth A B should be equal to the tension in the tie beam (at the joint) divided by the breadth of the tie beam multiplied by the value for C_1 given in Table XV.

Example I.—To show the application of these rules, we will apply them to joint 1 of the truss shown in Fig. 90, which is the truss shown by Fig. 60, for which the stresses were determined in Example II, page 89, and the sizes of the members computed in Example I, page 110, the sizes of the members being given in Fig. 90. The stress in the tie beam was found to be 16,260 pounds and the stress in the bottom of the rafter 21,300 pounds. The breadth of the tie beam is 6 inches, and the wood is to be white pine. The distance X, Fig. 81, should therefore $= \dfrac{16,260}{6 \times 80} = 34$ inches. The depth of the notch A B should $= \dfrac{16,260}{6 \times 1000}$, or $2\frac{3}{4}$ inches; this would give the proportions shown in Fig. 91. Even though the distance X may be great enough to resist the thrust of the rafters, a bolt should still be used, as shown in Fig. 91, to stiffen the joint and to prevent the rafter getting out of the notch.

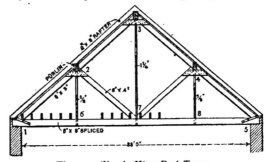

Fig. 90.—Simple King-Rod Truss.

Such a bolt also does much to prevent twisting of the timber while seasoning. The bolt need not be more than $\frac{1}{2}$ or $\frac{3}{4}$ inch in diameter.

There is still another very important point which needs to be considered when the foot of the rafter comes on the tie beam inside of the support, and that is the cross breaking of the tie beam. At the point of support the tie beam has to resist a vertical force equal to one-half the load on the truss—that is, when the truss is symmetrically loaded. This vertical force is considered as applied at the support. The tendency of this force is to shear the beam vertically across the grain, and also to break the beam at the toe of the rafter by cross bending, as shown by the dotted lines in Fig. 91. When, therefore, the toe of the rafter comes inside of the support the minimum depth of the tie beam on the line B E should be computed, both for shearing and bending.

Rule 3.—*The minimum depth of the tie beam at B E of Fig. 91 to resist shearing* should equal one-half the load on the truss, divided by the breadth of the beam multiplied by the value for F_s, Table XV.

APPLICATION OF RULE 3.

One-half of the load on the truss shown in Fig. 90 is 13,736 pounds (see Fig. 65) ; hence the depth of D E should equal $\dfrac{13,736}{6 \times 500}$ or 4⅝ inches. As the depth required for the notch was found to be

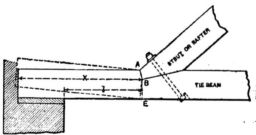

Fig. 91.—Showing Foot of Rafter Resting on Tie Beam Inside of the Support.

2¾ inches the depth B E will be 5¼ inches, if the beam is full 8 inches deep, or a little more than required by the formula.

Rule 4.—*To find the depth of the beam at B E necessary to resist the bending moment,* multiply one-half the load in pounds by the distance l, Fig. 91, in inches, and divide by three times the

breadth of the beam multiplied by the value for A, Table XV, and
extract the square root of the quotient: or, if we let $b =$ breadth
of the beam and $P =$ one-half of the load on the truss, then B E
should equal $\sqrt{\dfrac{P \times l}{3 \times b \times A.}}$

For the joint shown in Fig. 91 l measures 21 inches, therefore
B E should equal $\sqrt{\dfrac{13,736 \times 21}{3 \times 6 \times 60}} = \sqrt{267}$, or $16\frac{3}{4}$ inches. It is
therefore evident that we must reduce the distance X, so that the
joint will come over or nearer to the wall, and depend upon the
bolt to resist the thrust of the rafter, or else put a heavy bracket
under the truss. For a depth at B E of 5 inches it would be necessary to bring the point E within 2 inches of the support to resist
the cross strain.

For the truss shown in Fig. 90 the requirements of the construction of the cornice and the supporting of the rafters on the
purlin require that the joint shall be made as shown in Fig. 92.
With this construction it is impossible to have a shoulder more
than about 5 inches long, therefore the bolt must be made strong
enough to resist the entire stress in the strut. As the shoulder is
not expected to resist the thrust it is not necessary to make the
notch as deep as required by Rule 2, and, in fact, it is not necessary to make any notch at all, but for convenience in putting the
truss together it is a good idea to have a notch of about 2 inches in
depth. In the joint shown in Fig. 92 the center of the joint comes
above the wall, so that there is no tendency to shear the beam and
no bending moment. The only computation required by this joint,
therefore, is that for the size of the bolt.

Rule 5.—*To find the stress or tension in the bolt or strap when
used, as in Figs. 81, 82, 83 and 85*, draw to a scale of pounds the
line T, Fig. 92, parallel to the truss rafter and of a length equal to
the stress in the rafter. From the lower end of this line draw an
indefinite line parallel to the bolt or to the strap, and from the upper
end of the line T a line at right angles, or square to the seat of the
rafter, the two lines intersecting at the point c; then the line $b \ c$,

measured by the scale used in drawing the line T, will give the tension in the bolt or strap. Having found the stress the size of the bolt can readily be determined by Table XVI.

Table XVI.—Safe Strength of Bolts, in Pounds, for Stresses Computed as Explained under Rule 5.*

Diam.	Strength.	Diam.	Strength.	Diam.	Strength.	Diam.	Strength.
¾	6,000	1¼	17,700	1¾	35,000	2½	74,000
⅞	8,400	1⅜	21,400	1⅞	40,400	2¾	92,000
1	10,860	1½	25,740	2	45,000	3	108,000
1⅛	13,720	1⅝	30,000	2¼	60,000	3¼	130,000

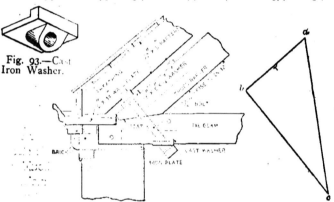

Fig. 93.—Cast Iron Washer.

Fig. 92.—Construction of Joint for Truss Shown in Fig. 90, when Rafters are Supported on Purlin.

From an inspection of the diagram, *a b c* in Fig. 92, it can readily be seen that the less the inclination of the bolt from a horizontal line the shorter will be the line *b c*, and therefore the smaller will be the stress in the bolt. For the joint under consideration the stress in the rafter is 21,300 pounds, and the line *b c* measures 34,800 pounds. From Table XVI we find that this will require a 1¾-inch bolt. On the underside of the tie beam a cast-iron washer, made as shown in Fig. 93, with a bearing for the bolt

*The values in this table are based on a unit stress of 20,000 lbs. per sq. inch. These values should not be used for rods, or for bolts subject to a direct stress.

head at right angles to the line of the bolt, should be used, as shown in Fig. 92. Without this washer it would be necessary to use at least a 4½-inch round cast-iron washer, and cut it into the tie beam, which would dangerously weaken the strength of the beam. If one does not wish to go to the expense of having this cast-iron washer made, a block of wood, preferably of oak or hard pine, may be bolted to the underside of the tie beam, as shown in Fig. 94. To prevent this block from slipping on the tie beam

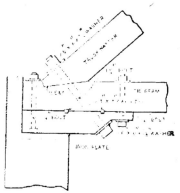

Fig. 94.—Block of Wood Used In stead of Cast Iron Washer.

square keys of oak should be driven into notches cut in the beam, as shown. Under the nut on the upper side of the rafter a square wrought iron or steel washer, the full width of the rafter, should be used. The pressure of the rafter against this washer must obviously be equal to the tension in the bolt, consequently the washer should be as large as practicable to prevent crushing the wood. The above is the only safe method of calculating the strength of a joint like that shown in Fig. 92.

PIN AND STRAP JOINTS AS SHOWN IN FIG. 95.

The points that must be considered in designing a joint like that shown in Fig. 95, are the strength of the strap, the bearing of the strap on the bolt, the bearing of the bolt in the timber and the resistance of the bolt to shearing.

The first step is to determine the tension in the strap. This must be found by means of a stress diagram, as explained under Rule 5. In order to reduce this stress as much as possible and also to keep the bolt back where it will not too much weaken the tie beam, it will be better not to place the strap at right angles to the rafter, but to give it an inclination about as shown in Fig. 95, the end of the rafter being cut at right angles to the center line of the strap. It is evident that the strap pulling on the bolt, which in this case acts as a pin, will cause the bolt to bear against the tie beam with a pressure equal to the stress in the strap. When we find the stress in the strap, therefore, we also have the pressure exerted by the bolt on the wood.

Table XVII.—Strength of Pin Bolts in Tie Beams When Used as in Fig. 95.

Dia. of bolt in ins.	Breadth of tie beam	Safe Strength in Pounds							
		In yellow pine		In Oregon pine		In spruce		In common pine	
1	6 inches	15,700		15,700		15,700		15,700	
1¼		24,540		24,540		24,000		21,000	
1⅜		29,680		29,680		26,400		23,100	
1½		35,340		32,400		28,800		25,200	⅝″ strap
1¾		42,000	⅝″ strap	37,800	⅝″ strap	33,600	½″ strap	29,400	
2		48,000		43,200		38,400		33,600	
2¼		54,000		48,600		43,200		37,800	
1¼	8 inches	24,540		24,540		24,540		24,540	
1⅜		29,680		29,680		29,680		29,680	
1½		35,340		35,340		35,340		33,600	
1⅝		41,160		41,160		41,460		36,400	
1¾		48,100		48,100		44,800		39,200	
2		62,832	1 3/16″	57,600		51,200	⅝″	44,800	⅝″
2¼		72,000		64,800		57,600		50,400	
2½		80,000		72,000	¾″	64,000		56,000	
1⅜	10 inches	29,680		29,680		29,680		29,680	
1½		35,340	¾″	35,340		35,340		35,340	
1⅝		41,160		41,160		41,460		41,100	
1¾		48,100		48,100		48,100		48,100	
2		62,800		62,800		62,800		56,000	¾″
2¼		79,500	1″	79,500	1″	72,000	13/16″	63,000	
2½		98,000		90,000		80,000		70,000	
1½	12 inches	35,340	¾″	35,340		35,300		35,340	
1¾		48,100		48,100	⅞″	48,100		48,100	
2		62,800		62,800		62,800		67,200	
2¼		79,500	1″	79,400		79,500		75,600	⅞″
2½		98,000		98,400		96,000	1″	84,000	
2¾		118,800	1 3/16″	118,800	1 1/16″	105,000		92,400	
3		41,400		129,400		115,400		100,800	

Computed for a bearing of 4,000 pounds per square inch for yellow pine, 3,600 for Oregon pine, 3,200 for spruce, 2,800 for common pine, and 20,000 for the strap and single shear of 10,000 pounds per square inch.

Table XVII gives the safe strength for bolts of several diameters in beams 6, 8, 10 and 12 inches thick, which will cover all ordinary cases. To find the size of the bolt, therefore, it is only necessary to look in this table and select the bolt having a safe strength for the kind of wood and the given breadth of the tie beam equal to the stress in the strap.

The values given in the table take into account the resistance of the bolt to shearing, and also the bearing of the strap on the bolt, so that no calculation for the bolt is necessary other than to find the stress in the strap.

The sectional area of the strap should be determined by dividing the stress in the strap by 20,000 pounds; the result will be in

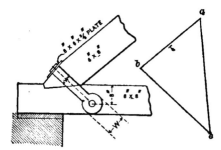

Fig. 95.—Joint with Strap and Pin.

square inches. The thickness of the strap should never be less than that given in the columns of Table XVII. The width of the strap should, of course, be equal to the sectional area divided by the thickness. The width of the head W should be equal to $1\frac{1}{3}$ w plus the diameter of the pin.

Example II.—To show the application of the above directions we will determine the size of the strap and bolt that would be required for the joint shown in Fig. 92, supposing that we wish to use a strap and pin in place of the bolt. The first step will be to locate the pin bolt and the center line of the strap, as in Fig. 95. Then draw the line T, Fig. 95, parallel to the rafter and equal to the stress in it, which in this case is 21,300 pounds, the same as in Example I, and from *b* draw a line parallel to the axis of the strap and from *a* a line at right angles to the seat of the rafter, these two lines intersecting at *c*. Measuring the line *b c* to the scale to which T was drawn, we find that it indicates 25,600 pounds, which is the stress in the strap. The sectional area of the strap should therefore be equal to $\dfrac{25,600}{20,000}$ or 1.28 square inches. From Table XVII we find that for a white pine tie beam with a breadth of 6 inches the strap must be $\frac{1}{2}$ inch thick. The breadth of the strap should therefore be equal to $\dfrac{1.28}{0.5}$ or 2.56 inches; or, say, 2⅝ inches. From the same table we see that a 1½-inch pin in a 6-inch white pine beam has a strength of 25,200 pounds, which is near enough to the stress. We should therefore use a 2⅝ x ½ inch wrought iron strap and a 1½-inch pin bolt. The width of the eye W should be 1⅜ x 2⅝ inches + 1½ inches, or 5 inches. Between the strap and the rafter we should place a 6 x 6 x ⅝ inch plate to prevent crushing of the wood. The joints shown in Figs. 92 and 95 should have about the same strength, and would probably cost about the same. The writer prefers the bolt joint, however, for the reason that it permits of being tightened by the screwing up of the nut on the bolt, while the strap joint can only be tightened by driving in steel wedges under the strap. Another objection to the strap joint is that it is somewhat difficult to bore the hole for the bolt exactly square to the tie beam, and if the hole is not true the strap will not bear evenly on the rafter. The strap joint, however, possesses the advantage that there is no projection on the bottom of the tie beam, which is a consideration when the bottom of the tie beam is to be cased.

Example III.—As a further illustration of the method of designing the joint at the foot of the principal strut we will take joint 1 of the truss shown in Fig. 69, and for which the loads and stresses are shown by Fig. 70. Fig. 96 shows the way in which the strut comes on the tie beam and the way in which the latter sets on the wall. For this joint a single bolt, used as in Fig. 92, will probably be sufficient. The problem is to determine the size

of the bolt. The stress in the rafter is 33,450 pounds. We find
the stress in the bolt by drawing the line *a b*, Fig. 96, parallel to
the rafter and equal to 33,450 pounds.

The line *a b* should be drawn to a scale of about 10,000 or
12,000 pounds to the inch. If one has not an engineer's scale,
an architect's scale of an inch to the foot can be conveniently used.

From *b* draw a line at right angles to the line *a b*, and from *a*
draw a line at right angles to the seat of the rafter, intersecting
the line from *b* at *c*. The line *b c* measured by the scale of T
equals 70,000 pounds. Therefore, if we use a single bolt, it must
be capable of resisting a stress of this amount, or, if we wish to
use two bolts, each bolt must be capable of resisting a stress of
35,000 pounds. From Table XVI we see that the safe load for a

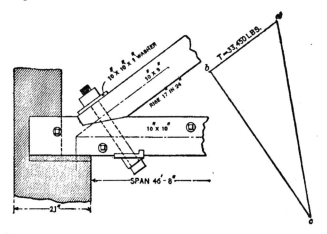

Fig. 96.—Detail of Joint 1,

2½-inch bolt is 74,000 pounds, and as a 2⅜-inch bolt would be less
than 70,000 pounds, we should use either one 2½-inch bolt or two
1¾-inch bolts. As the single bolt will weaken the tie beam and
rafter much less than two 1¾-inch bolts, it would be better to use
the single bolt. The joint therefore should be made as shown in
Fig. 96.

JOINTS AS IN FIG. 97.

When the rafter has an inclination of about 30 degrees and the stress in the rafter exceeds 30,000 pounds a joint similar to that shown in Fig. 97 is generally the best to use. In this joint the end of the tie beam and the plate P offer a good deal of resistance to the thrust of the strut, even without any assistance from the bolt. As will be seen, the rafter is held in place by the plate P, and before this plate can be moved it is necessary either to shear off the top of the tie beam beyond the point B or to break off the lugs on the plate P.

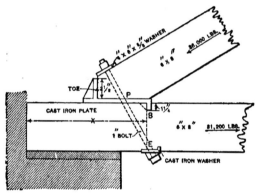

Fig. 97.—Joint for Rafter with Inclination of 30 degrees or less and Stress in Excess of 30,000 Pounds.

If the distance X is long enough to resist the shearing tendency then a single bolt may be used of about 1 or 1¼ inches in diameter merely to hold the rafter in place and provide against any possible tendency of sudden failure. Even when the distance X is not great enough to resist the entire shearing stress, it would still be great enough to resist a considerable portion, so that in the opinion of the writer if the bolt is made strong enough to resist one-half of the stress in the rafter (which would usually cover the actual dead load on the truss) the joint might be considered as perfectly safe.

Example IV.—As an example of designing a joint of this kind we will assume that the rafter in Fig. 97 has a stress of 36,000 pounds; the corresponding tension in the tie beam would be 31,200 pounds and the supporting force 18,000 pounds. The wood to be yellow pine. The position of the joint in the wall to be as shown in the figure. This gives us the length for X of 22 inches. To find the length for X required to resist the entire thrust in the rafter we use Rule 1, page 122, which is to divide the stress in the tie beam by the breadth of the tie beam, multiplied by the resistance of the wood to shearing. In this case we should use the values for F_1, in Table XV, because the wood is under compression and tests have shown that under such conditions the resistance of the tie beam to the shearing is greatly increased. Dividing the stress in the tie beam, 31,200 pounds, by the product of the breadth by the value of F_1 for yellow pine, we have 21 inches as the length required. As the actual length of X is 22 inches, the tie beam is capable of resisting the thrust without any assistance from the bolt, and for this example a 1-inch bolt is all that is needed.

HIGHT OF THE TOE (FIG. 97).

If we are to depend upon the resistance of the tie beam to shearing for the strength of the joint, then the hight of the toe must be sufficient to prevent the end of the rafter from crushing the wood. The hight of the toe should therefore be equal to the stress in the tie beam, divided by the breadth of the rafter, multiplied by the value for C_1, given in Table XV, which for this example would be $\dfrac{31,200}{6 \times 1500}$, or $3\frac{1}{2}$ inches; therefore the depth of the toe should be $3\frac{1}{2}$ inches. The plate P should be made of cast iron, at least $\frac{3}{4}$ inch thick, and should have a bracket back of the lug at the toe to prevent its being broken. The lug at B should also be strengthened by a bracket, as shown by dotted line. The lug at B has to resist a stress equal to that exerted by the toe of the rafter, and one might think that it should be let into the tie beam to a depth equal to the distance found for the hight of the toe. There are three reasons, however, why the lug should not be as deep as the toe.

1. If there were no friction between the plate and the beam and none between the strut and the plate, the compression against both lugs would be the same, but the element of friction is always present and is greatest when the stress in the strut is the greatest, and this friction will very materially reduce the stress on the lug B.

2. Lug B bears square on the end fibers of the tie, and hence a greater unit stress may be permitted than in the case of the toe, where the pressure is at an angle with the fibers.

3. Too deep a cut at B would seriously affect the strength of the tie beam, so that it is desirable to make it as little as possible, while the toe can be made a little larger than might be absolutely necessary without doing any harm.

When making a joint of this kind one should be careful to see that the net sectional area of the tie beam at B E is sufficient to resist the tensile stress. In this particular case the net sectional area required will be equal to $\dfrac{31,200^*}{2000}$ or not quite 16 square inches; and as the breadth of the beam is 6 inches this would require a clear depth between B and E of about 3 inches. The actual depth is nearly 6 inches, consequently the beam has sufficient strength at this point. For trusses in which the stresses run up to 60,000 or 70,000 pounds a joint like that shown in Fig. 84 should be used, the bolts being made strong enough to resist one-half of the thrust in the rafter and the tie beam being depended upon to resist the balance. With these examples the reader should be able to choose the best form of end joint and to proportion the parts with absolute safety.

BRACING OF TRUSSES WHEN SUPPORTED BY POSTS.

When a truss is supported by posts a rigid connection is necessary between the post and the truss to prevent racking of the building by wind pressure. This connection is generally made

---
*See Rule 1, page 107.

by means of a brace, which should be secured at the ends so that
it may act either as a tie or as a strut. Fig. 98 is a good example
of the end construction of a heavy wooden truss supported by

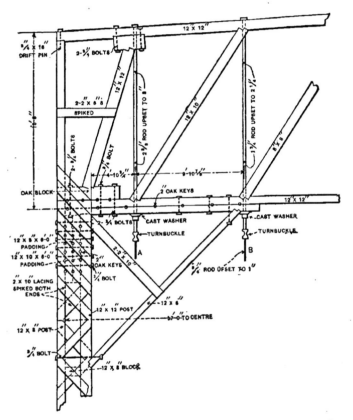

Fig. 98.—Good Example of End Construction of Heavy Wood Truss Supported by Wooden Posts.

wooden posts. This represents a portion of one of the four trusses
which supported the domed roof and ceiling of the Music Hall
of the Pan-American Exposition. The length of the trusses was

about 114 feet. It should be noticed that the last two panels of
the lower chord are reinforced by a 6 x 12 inch piece keyed and
bolted to the underside. At one end this piece is notched over
the top of the inside post of the main column, at the other end
it receives the thrust of a knee brace to the same post. It should
also be noticed that the brace is bolted to both the tie beam and
the post. The manner in which the joints are made at the top
and bottom of the end strut of the truss is also of interest. Such
joints would be advisable only when the inclination of the strut
is 60 degrees or more. The rods A and B support a suspended
dome ceiling below the truss.

JOINT BETWEEN BRACE AND RAFTER.

The joint at 2, Fig. 90, should be made as shown in Fig. 99.
When the rod at this joint is less than 1 inch in diameter a notch
may be bored in the top of the rafter to form a bearing for the
washer, which may be a round cast-iron washer of the ordinary
pattern. The bottom of the brace should be slightly notched into
the bottom of the truss rafter, as shown in Fig. 99.

DEPTH OF THE NOTCH.

When the inclination of the brace and of rafter is 40 degrees
or more the notch for the brace may be made as in Fig. 99,
and not more than $\frac{3}{4}$ inch deep. When the inclination of the rafter
and brace is only about 30 degrees then a notch should be made
as shown in Fig. 100, and the depth of the notch should be proportioned to the horizontal projection of the stress in the brace
by the following rule:

Rule 6.—The depth of the toe d of any brace should be equal
to the horizontal pressure divided by the breadth of the brace,
multiplied by the value of C_1 in Table XV. The horizontal pressure will be equal to the horizontal projection of the stress in the
brace, which can always be readily obtained from the stress
diagram.

Example V.—To show the application of Rule 6 we will as-

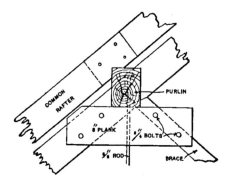

Fig. 99.—Detail of Joint 2 of Fig. 90.

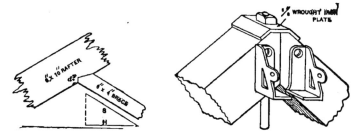

Fig. 100.—Showing Depth of Notch for Brace.

Fig. 101.—Detail of Joint 3 of Fig. 90.

Fig. 102.—Duplex Hanger with Trusses.

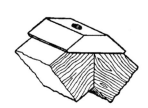

Fig. 103.—Cast Iron Shoe for Heavy Trusses.

sume that the stress in the brace, Fig. 100, is 8300 pounds; that the brace is of the dimensions shown and has an inclination of 30 degrees. If the line S, Fig. 100, represents the stress in the brace, then the line H will represent the horizontal pressure on the toe. The horizontal pressure corresponding with S equals 8300 pounds, at 30 degrees, is 7190 pounds. We will assume that the brace and rafter are of spruce, for which wood the value of C_1 is 1200 pounds. Then by Rule 6 the depth of the toe d should equal $\dfrac{7190}{6 \times 1200}$, or 1 inch.

The best manner of supporting the purlin is by means of duplex hangers let into the rafter. For beams 10 or 12 inches wide these hangers are made as shown in Fig. 101 ; for beams up to 6 inches in width they are made in one piece, with a single nipple, as shown by Fig. 102 ; and for beams 8 inches wide they are made in both styles. These hangers are easily applied; they furnish a connection fully as strong as the purlin, and tie the purlins together longitudinally. They are carried in stock in the larger cities.

When the load on the purlin is not very great a 3-inch plank may be bolted to each side of the rafter and brace to support the purlin, as shown in Fig. 99. It is always best, however, to use a duplex hanger when they can readily be obtained. The common rafters should be notched on the purlin, as shown in Fig. 99.

Joint 3 *of Fig.* 90.—This joint should be made as shown in Fig. 101, the peak of the joint being cut off and a bent plate of wrought iron or mild steel set over the top of the rafters to receive the nut on the rod. For very heavy trusses a cast-iron shoe, made as shown in Fig. 103, should be used in place of the bent plate. If the weight on the purlin is not very great it may be supported by a plank bolted to the rafters, as in Fig. 99.

Joint 7 *of Fig.* 90.—The best way of making this joint is shown in Fig. 104, the oak angle block being let into the tie beam about ¾ inch for a 6-inch beam and 1 inch for an 8-inch beam. The hight of the bevel d should be at least equal to that which would be obtained by Rule 6, and never less than 2½ inches. A tenon on the brace may be let into the tie beam, as shown, to keep the brace from moving sideways. The length, *l*, of the angle block should be at least equal to the result obtained by Rule 7.

Rule 7.—To find the least length, *l*, of the angle block, as in Fig. 104, divide the stress in the rod less the weight of the floor or ceiling at that point by the breadth of the beam, multiplied by the value for C_2, Table XV.

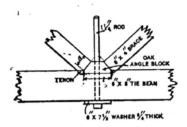

Fig. 104.—Detail of Joint 7 of Fig. 90.

Example VI.—In the truss shown in Fig. 90 the stress in the rod, less the ceiling load at joint 7, is 9100 pounds; the breadth of the tie beam is 6 inches and the value for C_2 for white pine is 250 pounds, therefore the length *l* should equal $\dfrac{9100}{6 \times 250}$, or 6 inches. The block, however, should be about 7½ inches long to give a good proportion to the joint.

JOINT FORMED BY PRINCIPAL STRUT AND TOP CHORD.

A joint such as joint 3 of Fig. 66 should be constructed as shown in Fig. 105, a slight offset being made at the middle of the joint to hold the top chord until the load is brought on the truss. Sometimes a bent wrought-iron plate is used for the washer, but a cast-iron shoe, like the one shown in Fig. 105, is better. It is sometimes desirable to extend the top chord beyond the head of the strut, as in Fig. 106 and also in Fig. 79. When this is done a plain rectangular washer is sufficient. The depth of the notch *d* should always be proportioned by Rule 6. Sometimes two notches are made to receive the end of the brace, as shown by the dotted lines in Fig. 106.

In the opinion of the writer this does not make as good a joint as the single notch, for the reason that it is very difficult to fit the end of the strut so that it will bear evenly at both places, while with a single notch the full bearing must necessarily be brought on the toe. Also, when there is a double notch the small triangular piece between the notches will generally offer but little resistance to shearing. The only place where two notches would be at all desirable is where the inclination of the strut is very steep, as is the case of the struts in Fig. 98.

In Fig. 106 is also shown the ordinary method of making a joint between any angle strut and the top chord, and if the toe d is proportioned by Rule 6 it will answer for the ordinary roof trusses. If there is a counter brace at the joint it may cut against the main brace and the top chord, as shown by the dotted lines. Fig. 106 reversed will generally answer for the joint between the tie beam and any main brace, as at joints 3 and 5 of Fig. 79. The notch for the toe of the brace, however, should never be made greater than required by Rule 6. When the stress in the brace is very great, so as to require a deep notch for the toe, it would be better to use an angle block cut into the tie beam for the strut to bear against, as in Fig. 104, as it is not necessary to make so deep a notch for the angle block. Wherever a brace and counter brace come together, either at the top chord or tie beam, they should bear against an angle block, as shown in Fig. 104, as the counter brace is not always under load.

WASHERS.

The rods in wooden trusses should always have a washer under the head or nut, proportioned so that the tension in the rod will not crush the wood. The resistance of the common framing woods to crushing against the grain without indentation is given in the column headed C_2, Table XV.

Rule 8.—To find the area of the washer divide the total stress in the rod by the value for C_2, Table I.

Example VII.—The total stress in the king rod of the truss shown in Fig. 90, which is also the rod shown in Fig. 104, is 11,089 pounds. Dividing this by 250, the value of C for white pine, we have 44 square inches as the area of the washer. As the tie beam is only 6 inches wide the washer should be 6 x 7½ inches.

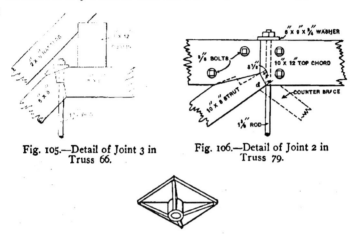

Fig. 105.—Detail of Joint 3 in   Fig. 106.—Detail of Joint 2 in
        Truss 66.   Truss 79.

Fig. 107.—Cast Iron Washer for Large Rods.

Example VIII.—Rod A of Fig. 108 has a total stress of 13,490 pounds. As the wood is white pine, the area of the washer should be 13,490 divided by 250, or 54 square inches, which is equivalent to 6 x 9 inches, the long way of the washer being placed crossways of the tie beam. Rod B has a total stress of 5800 pounds, which requires 23 square inches, or 4 x 6 inches, for the size of the washer. Rod C has a stress of 1908 pounds, which requires a washer of only about 8 square inches, but as this would give a very small bearing for the tie beam the washer should be made at least 6 x 4 inches. When the washer projects more than about 3 inches beyond the head of the nut a cast-iron washer with ribs, as shown in Fig. 107, should be used.

Table XVIII gives the maximum stress which round and rectangular washers will resist without sinking into the wood. To use this table, look for the stress in the table under the kind of wood that is to be used nearest or just above the stress in the rod, and the required size of washer will be found on the same line in Column 1.

Thus, for a rod subject to a stress of 13,490 pounds, the wood being white pine, we see that a 6 x 9 washer will be required, as found in Example VIII.

The diameters of the round washers are those of the standard sizes of cast-iron washers given in Table XIX. Comparing the values given in Table XVIII for round washers with the strength of the rods for which they are intended, it will be seen that the bearing resistance of the washers on white pine and spruce is only about one-half of the working strength of the rod, consequently for white pine, spruce and Oregon pine the standard size of washers is not large enough for the strength of the rod. As a rule, for the rods of wooden trusses it is best to use rectangular washers cut from steel plates, cutting the washers to the required size. It is, of course, not really dangerous to use smaller washers than would be required by Table XVIII, as a little crushing of the timber will not endanger the safety of the truss, but it is best to keep within the limits of Table XVIII when practicable.

Very large washers should be made of cast iron with brackets, as in Fig. 107.

Table XVIII.—Safe Bearing Resistance of Washers in Pounds.

| | ROUND WASHERS. | | | |
Diameter.	White Pine and Spruce.	Oregon Pine.	Georgia Pine.	Oak.
2⅝	1,350	2,160	2,700	3,240
3	1,760	2,820	3,520	4,230
3¼	2,070	3,300	4,140	4,970
3¾	2,760	4,400	5,520	6,620
4	3,140	5,020	6,280	7,530
4¾	4,430	7,080	8,860	10,630
6	7,060	11,400	14,100	16,960
6¼	7,660	12,260	15,300	18,400
7¼	10,300	16,500	20,600	24,700
8¼	12,900	20,700	25,800	31,100
9¼	16,250	26,000	32,500	39,000
10¼	20,000	32,000	40,000	48,000

RECTANGULAR WASHERS.

Size of Washer, ins.	White Pine and Spruce.	Oregon Pine.	Georgia Pine.	Oak.
4 × 6	6,000	9,600	12,000	14,400
4 × 8	8,000	12,800	16,000	19,200
6 × 6	9,000	14,400	18,000	21,600
6 × 7	10,500	16,800	21,000	25,200
6 × 8	12,000	19,200	24,000	28,800
6 × 9	13,500	21,600	27,000	32,400
6 × 10	15,000	24,000	30,000	36,000
8 × 8	16,000	25,600	32,000	38,400
8 × 9	18,000	28,800	36,000	43,200
8 × 10	20,000	32,000	40,000	48,000
8 × 12	24,000	38,400	48,000	57,600
10 × 10	25,000	40,000	50,000	60,000
10 × 11	27,500	44,000	55,000	66,000
10 × 12	30,000	48,000	60,000	72,000
10 × 14	35,000	56,000	70,000	84,000
12 × 12	36,000	57,600	72,000	86,400
12 × 14	42,000	67,200	84,000	100,800
12 × 16	48,000	76,800	96,000	115,200

Table XIX.—Proportions of Standard Cast Washers.

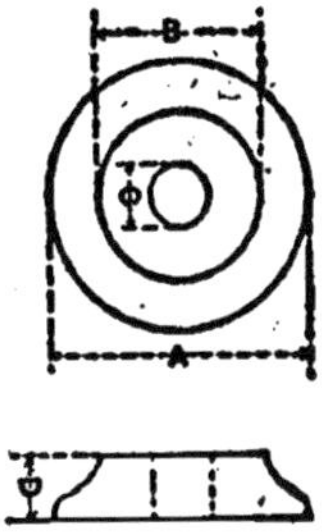

For sizes not gives below.
Diameter of bolt = d.
$A = 4d + \frac{1}{4}''$. $C = 1d + \frac{1}{8}''$.
$B = 2d + \frac{1}{4}''$. $D = 1d$.
all dimensions in inches.

Standard Cast Washer.

Diameter of Bolt = d.	A.	B.	C.	D.	Weight in Pounds.
½	2⅝	1¾	9/16	⅝	½
⅝	3	1⅞	11/16	¾	¾
¾	3¼	2⅛	13/16	⅞	1¼
⅞	3¾	2½	15/16	⅞	1½
1	4	2¾	1 1/16	1⅛	2½
1⅛	4¾	2¾	1 3/16	1⅛	3
1¼	6	3	1 5/16	1⅜	5¾
1½	6¼	3¼	1⅝	1½	6
1¾	7¼	3¾	1⅞	1¾	9½
2	8¼	4¼	2⅛	2	17¼
2¼	9¼	4¾	2⅜	2¼	20
2½	10¼	5¼	2⅝	2½	27¼
2¾	11¼	5¾	2⅞	2¾	36
3	12¼	6¼	3⅛	3	46

The foregoing rules and illustrations cover all of the joints that occur in king, queen and Howe trusses, except the splicing of the tie beam, which we will now consider.

BUILT-UP TIE BEAMS.

When the tie beam of a roof truss cannot be made from a single stick of timber it will generally be better to build up the beam out of 2 or 3 inch plank, breaking joints and bolted together, rather than to try to splice solid timbers. In building up a tie beam of plank the planks must break joint in such a way that the net cross section, after deducting for joints and bolt holes, will be sufficient to resist the tensile strength in the tie. There must also be enough bolts between any two consecutive joints to transmit the stress from one layer to the next. To meet this requirement usually necessitates considerable study in laying out the tie beam.

Example IX.—The method of computing the number of bolts required can best be shown by an example. Fig. 108 represents a little more than one-half of the tie beam of the truss shown in Figs. 79 and 80. On page 116 we found the net sectional area required in the tie beam at the different panels to be as indicated in Fig. 108. We also concluded to build the tie beam with five 2 x 10 inch planks of white pine. The size of the braces, rods and washers to be as indicated in Fig. 108. The rods will, of course, pass through the center plank, consequently for the two end panels the center planks C C' might as well be jointed at the rods, for they would be pretty nearly cut in two, anyway. The center rod being only ¾ inch in diameter, we will make the center plank C″ in one piece from rod B to the corresponding rod on the other side of the truss. The outer layers of plank on each side should have but one joint if possible. In this case a 26-foot plank will extend beyond the center of the truss, so that for the outer layers, A and A', we will use two planks 26 feet long and two planks 23 feet 4 inches long, starting the 26-foot planks from opposite ends of the beam, so that the joints will come on opposite sides of the center, as shown at Y and Z of Fig. 108. The joints in the next two layers, B and B', should come as far from the center of the truss as the longest stock length will permit. We will therefore make the planks B and B' 28 feet long, piecing out at each end with planks 10 feet 8 inches long. Now if we look at the plan of the tie beam we will see that the planks A and B, if bolted together, form a continuous tie from one end of the truss to the other, and that opposite the joint X the net sectional area is that of two 2 x 10 inch planks, or 40 square inches. Therefore the planks B and B' will sustain all the tension in the two center panels of the tie beams, and the problem is to bolt planks A and B and A' and B' together so that the planks A and A' will receive the stress in the second panel from B and B' and transmit it to the end of the truss. We may consider that the plank B will have to transmit to A one-half of the stress in the second panel, which is 43,260 pounds, or 21,630 pounds for the plank B. We must therefore ascertain how many bolts are required between the points X and Y to transmit 21,630 pounds. This we do by means of Table XX, which gives the maximum stress it is safe to allow for ¾, ⅞ and 1 inch bolts in 2-inch planks.

Table XX.—Maximum Allowable Stress on Bolts in 2-Inch Plank of Built-up Tie Beams.

Diam. of bolt.	Yellow pine Stress. d.		Oregon pine. Stress. d.		Spruce. Stress. d.		White pine. Stress. d.	
¾ inch	2,250	5	2,024	4½	1,800	5	1,500	5
⅞ inch	2,620	6	2,366	5	2,100	5½	1,760	5½
1 inch	3,000	6	2,700	5½	2,400	6½	2,000	6

From this table we see that a ¾-inch bolt in white pine will transmit 1500 pounds. We should therefore require as many ¾-inch bolts between X and Y as 1500 is contained in 21,630, or 14 and a fraction. This would bring the bolts quite close together.

The number of ⅞-inch bolts required is equal to $\dfrac{21,630}{1760}$ or 12.

The number of 1-inch bolts required is equal to 21,630 divided by 2000, or 11 bolts. It will probably be best in this case to use 12 ⅞-inch bolts. The figures under *d* indicate the distance in inches

Fig. 108.—Detail of Tie Beam of Fig. 10.

that the center of the bolts nearest the joint should be from the joint. There should always be two bolts at each side of each joint, which will take 4 of the 12 bolts required, leaving 8 bolts to be spaced evenly between.

From Table XX we see that ⅜-inch bolts in white pine should be spaced 5½ inches from the joint, so that the distance between the pairs of bolts at X and Y is 11 feet 9 inches; and as there will be nine spaces this will make the bolts 1 foot 3⅝ inches on centers. These bolts should be staggered as shown in the elevation, Fig. 108. Five and one-half inches from joints Y and Z toward the center there should also be 2 more ⅜-inch bolts. Between the joint Y and the other end of B we must also place 12 ⅜-inch bolts, and from there to the right-hand end of the truss, and also from X to the left-hand end of the truss, we should place ¾-inch bolts about 2 feet on centers. These latter bolts are required to bind the planks together so as to give a solid and even bearing for the main strut, and also for the ceiling joist. For lighter trusses ⅜-inch bolts might be used for the end panels. To properly unite the planks of the tie beam, therefore, will require 14 ⅜-inch bolts and 7 ¾-inch bolts in each half of the truss.

The proper lengths of plank to be used in bolting up tie beams of different lengths will usually have to be studied out for each particular case, having in mind the longest length of planks that can readily be obtained. The principal aim should be to get as great a distance between consecutive joints, as X and Y, as possible, and not to have more than two joints opposite each other, or more than one joint opposite the rod besides the joint that is in the center plank. It should also be remembered that in Howe trusses the tension is always greatest at the center of the truss; in Howe trusses, therefore, there should be as few joints near the center of the beam as practicable. In queen trusses the tension is usually greatest in the end panels.

SPLICING AND BOLTING OF TOP CHORD.

As the top chord is always in compression, it is not necessary to be as particular with the splicing and bolting as with the tie beam. For the top chord it will be better to use 3 or 4 inch planks, if they can be obtained. The splices should be made near the joints, and the planks should be bolted together with ⅝ or ¾ inch bolts, spaced from 2 to 2 feet 4 inches on centers.

Thus for the top chord of the truss shown in Figs. 79 and 80, which must be 10 x 12 inches, we should use 4 x 12 inch planks for the center layer and 3 x 12 inch planks for the two outer layers, giving a total breadth of 10 inches. The 4-inch planks we would make in three lengths of about 16 feet each. The 3-inch planks should be jointed near joints 2 and 6.

Two bolts should always be placed at each side of each joint. Fig. 106 shows how the bolts should be placed at joint 2.

In making the drawing for a roof truss, first draw out a truss to fit the roof and ceiling, then compute the loads, draw the truss and stress diagrams and determine on the size of the members. Then detail the joints, as in these articles, and finally make a ½ or 1 inch scale drawing of the truss to conform with the proportions worked out for the joints and members.

JOINTS IN SCISSORS TRUSSES.

Scissors trusses are a distinct type of truss, and the joints in these trusses, because of the inclination of the members, differ from those in almost all other forms.

The typical forms of scissors trusses, and the way in which they are strained are described in Chapter IV. The scissors truss is susceptible to great variations, but the joints are similar in all of them. Some of the joints are very easy to make, but there are always two, and sometimes three or four, which in trusses of wide span, or in trusses which support large roof areas, must be carefully proportioned in order to successfully resist the stresses.

The trusses shown in Figs. 109, 110, 113, 115 and 117 have been selected as representative of the most common forms of scissors trusses, and embracing the different kinds of joints which have to be made.

Fig. 109 is about the simplest form of the scissors truss, and the construction shown is also the simplest. This truss was used by the writer over a church in a small town, where rods and special iron work would have been expensive. As indicated by the en-

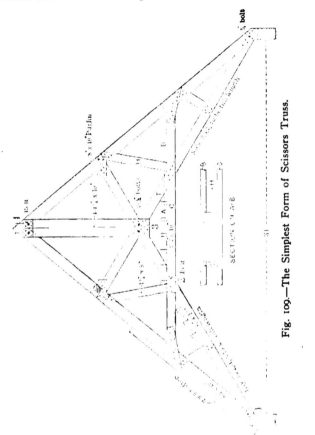

Fig. 109.—The Simplest Form of Scissors Truss.

graving, the truss was built of 1-inch rough spruce boards, spiked together, and the joints were formed by lapping and spiking the boards. The joints in this truss, which require special attention, are those numbered 1, 2, 3 and 4. The joint at 1 must

be capable of transmitting almost the entire thrust in the rafter
to the tie beam. In this case the boards of the tie beam and rafter
interlock, and are well spiked together, and the pieces are further
secured by one ¾-inch bolt.

Four boards in each tie beam are the full length, so that one-
half of the sectional area is available for transmitting the tensile
stress, and the other half is used principally for filling, although
it assists in resisting the transverse train produced by the ceiling
joists.

Joints 2 and 3 are also almost entirely in tension and require a
good many spikes. The cross tie C was made of two planks, one
on each side of the tie beams, so that they do not weaken the latter.
The main vertical tie was made of four boards, in order to get
a large area for nailing. This vertical tie always has a great stress
(see Fig. 76), and when made of boards or plank it requires a
large number of spikes to hold the ends.

In all of the joints 1, 2, 3 and 4, there should be at least one
¾-inch bolt, in addition to the spikes, to prevent the outer boards
from curling or separating as they season.

The author does not consider this construction as good as that
shown in Fig. 110, but in country towns it may be safely used
for spans of 34 feet, and with trusses not over 12 feet apart, provided that the inclination of the rafter and tie beams is about as
shown in Fig. 109. When the angle between the rafter and tie
beam is greater than in the figure, the boards do not lap on each
other as much, so that there is not as much room for spikes. On
the other hand, it is not desirable to have a sharper angle at 1,
as the sharper the angle the greater will be the stress in the tie
beam.

With this form of construction, cement-coated spikes should
be used if they can be obtained, as they "hold" much better than
the common spikes.

The truss indicated in Fig. 110 represents truss B in the half-
tone view, Fig. 120, and shows the best construction for trusses
of this type. The only joints in this truss which require any special attention are those at A, A. To prevent any spreading of

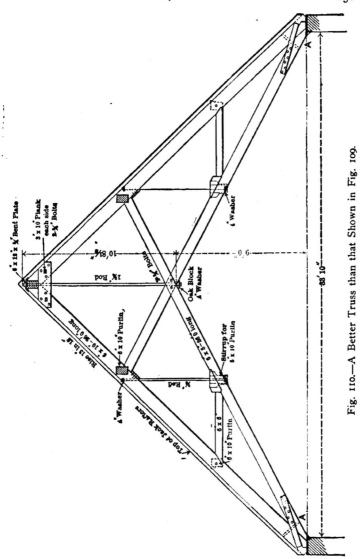

Fig. 110.—A Better Truss than that Shown in Fig. 109.

the rafters under the greatest possible load on the roof will produce a stress in the tie beam of this truss of about 25,000 pounds,
hence the joints at A, A, must be capable of resisting 25,000
pounds each.

The author has found that the best method of making this joint
is by means of a wrought-iron strap and lag screws, as shown
by the enlarged detail, Fig. 111. Lag screws are preferable to
bolts, for the reason that it is almost impossible to get the holes
in the strap and in the wood in line, and usually the hole in the
wood has to be made so large that the bolt does not fit tightly.
With lag screws each screw is bound to get a good bearing in
the wood. The holes in the two sides of the strap must, of course,
be staggered, so that they will not come opposite to each other.

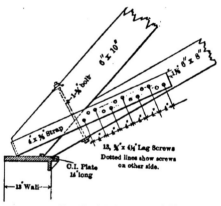

Fig. 111.—Detail of Joint at A of Fig. 110.

The net sectional area of the strap should be at least equal to
the stress in the tie beam divided by 20,000 pounds.

The number of lag screws (for both sides) is found by dividing the stress in the tie beam by the resistance of one screw. The
author has been unable to find any data on the resistance of lag
screws, but believes that the following rules may be used with
safety :

Safe Resistance of Lag Screws, When Used as in Fig. 111.

½ x 4   inch lag screw, 1,000 pounds, minimum thickness of strap, ¼.

⅝ x 4   inch lag screw, 1,500 pounds, minimum thickness of strap, ¼.

¾ x 4½ inch lag screw, 2,000 pounds, minimum thickness of strap, 5-16.

⅞ x 5   inch lag screw, 3,000 pounds, minimum thickness of strap, 5-16.

For the truss shown in Fig. 110 the above rules would require a net sectional area, in the strap, of $\dfrac{25,000}{20,000}$ or 1¼ square inches, and the number of ¾-inch lag screws, $\dfrac{25,000}{2,000}$ or 13.

With a thickness of ⅜-inch, the width of the strap must equal $\dfrac{1.25}{\frac{3}{8}}$, or 3⅓ inches, and to this must be added the width of one hole, making the total width 4 inches. In addition to the strap, the tie beam and rafter should also be bolted together by one bolt, to bring them tightly together before the strap is put on, and also to hold them together when the truss is raised.

The other joints of this truss require no computations.

The joint at the intersection of the tie beams is best made by halving the timbers and bolting them together.

The roof purlins in this roof were hung in Duplex hangers, which, in the opinion of the author, offer the most convenient and best means of support.

Another method of joining the rafters and tie beams of scissors trusses is shown by Fig. 112, in which the horizontal thrust of the rafter is resisted by a single bolt. When the inclination of the rafter is quite flat, this joint is to be preferred to that shown by Fig. 111. It also has the advantage, that where the truss is erected, one piece at a time, the tie beams may be put up first, and a seat is provided to receive the rafters. The strap prevents the end of the rafter from springing up.

When the center of the ceiling is to be level, a truss with a horizontal tie, as in Figs. 113 and 115, is sometimes more economical than one of the shape shown by Fig. 110, as the horizontal tie permits of the use of smaller rods, and also affords a support for the ceiling joists. When the conditions will admit, it is better to make the horizontal tie of two planks, and bolt them to the tie beam (one on each side) by a single large bolt, as in Fig. 114.

If the horizontal tie must be kept flush with the inclined ties,

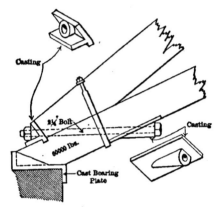

Fig. 112.—Another Method of Joining Rafters and Tie Beams.

as will usually be the case when the ties are exposed, then the joint should be made as shown in Fig. 116.

The diameter required for the bolt in Fig. 114 may be deter-

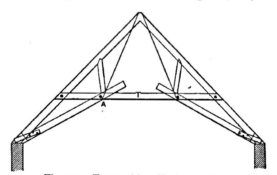

Fig. 113.—Truss with a Horizontal Tie.

mined by means of Table XXI, which gives the maximum safe resistance for one bolt without crushing the wood. The bolt should have a resistance equal to the full stress in the horizontal tie; thus if the stress in the latter is 22,000 pounds it will require a 2½-inch bolt in the joint.

It is always better to use one single bolt (up to about 3 inches in diameter) than two smaller bolts, for the reason that where two bolts are used, one bolt is almost sure to be strained more than the other.

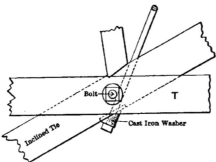

Fig. 114.—Detail of Joint at A of Fig. 113.

Table XXI.—Maximum Allowable Stress on Bolts Used as in Fig. 114.

Diameter of bolts in inches.	Thickness of planks, T. Inches.	Yellow pine. Pounds.	Oregon pine Pounds.	Spruce. Pounds.
1	2	4,400	4,400	4,400
	3	3,000	3,000	3,000
1¼	2	7,500	7,500	6,000
	3	5,720	5,720	5,720
1½	2	9,000	8,000	7,200
	3	10,000	10,000	10,000
1¾	2	10,500	9,400	8,400
	3	15,600	14,000	12,600
2	2	12,000	10,800	9,600
	3	18,000	16,200	14,400
2¼	2	13,500	12,000	10,800
	3	20,200	18,000	16,200
2½	2	15,000	13,500	12,000
	3	22,500	20,250	18,000

When the stress in the horizontal tie is very great, as is the case in a truss like that shown in Fig. 117, it is difficult to make a satisfactory connection with a wooden tie, and it is much better construction to use two rods, as shown in Figs. 117 and 118. By using a cast-iron washer, as shown in Fig. 118, a very strong joint is easily made.

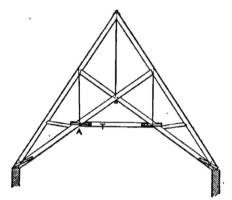

Fig. 115.—Another Example of Truss with Horizontal Tie.

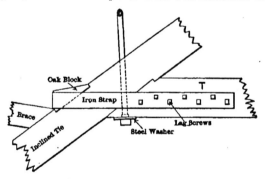

Fig. 116.—Form of Joint Where Horizontal Tie is Kept Flush with the Inclined Ties.

Wherever rods or large bolts intersect a tie at an angle other than 45 degrees cast-iron washers having a seat for the nut at right angles to the axis of the rod (or oak blocks) should always be used, as otherwise it will be necessary to cut such a large hole in the tie to get a sufficient bearing for the washer as to very seriously weaken it.

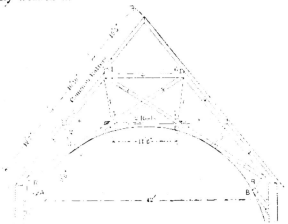

Fig. 117.—One of the Common Forms of Scissors Truss.

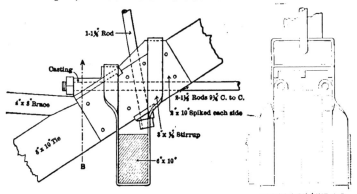

BECTION ON LINE A-B.

Fig. 118.—Details of Joint Where Two Rods Are Used in Connection with Wooden Tie.

The foregoing details cover those joints in scissors trusses which are peculiar and difficult to make.

There is one other connection, however, which it may be well to mention, and that is where two diagonal trusses intersect at the center. When a church has three or four gables, or arms roofed by trusses, it is generally necessary to support the roof over the "crossing" by means of diagonal trusses.

If the trusses are of the shape shown in Fig. 117, two full trusses may be used, and the rods and top chord arranged so that those of one truss will pass above those of the other, and in that case each truss may be figured to support one half of the roof.

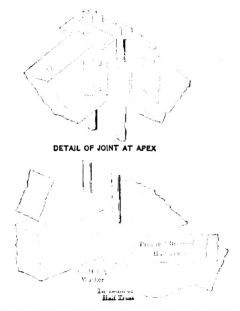

DETAIL OF JOINT AT APEX

Fig. 119.—Details of Center Joints of Diagonal Truss D D, Fig. 120, Showing Manner of Supporting the Half Truss.

If either of the forms of truss shown in Figs. 110, 113 or 115 are used, however, it is impracticable to pass the trusses by each other, or to join them at the center, so as to obtain the strength of two trusses. In such cases the author is in the habit of building one diagonal truss capable of supporting the entire roof load, and suspending two half trusses at its center. Fig. 119 represents the

top and bottom joints at the center of a diagonal scissors truss
used by the author and shown in Fig. 120. The photographic
view shows the whole diagonal truss, D D, in place, and supporting one half truss behind it. The other half diagonal had not been
set when the photograph was made. In addition to the diagonal
trusses, three trusses like Fig. 110 were used, one of which is
shown at B. Instead of using a single rod at the center of the

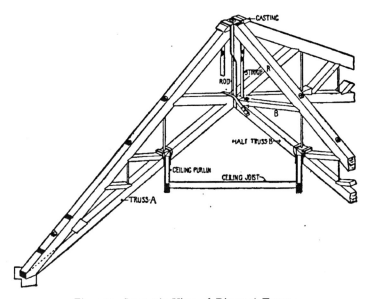

Fig. 122.—Isometric View of Diagonal Trusses.

through truss, two rods were used, one on each side of the truss,
and the washer at the bottom extended 4 inches beyond the sides
of the truss, to receive the abutting half trusses. The tie beams
of the half trusses were also tied together by two bent straps,
spiked to each side of each tie beam and passing over the tie beams
of the through truss.

The tops of the half trusses were secured to the top of the

Fig. 120.—Showing Diagonal Truss Over a Church Having Three Arms
of Equal Width and Three Gables.

Fig. 121.—View of Church After Completion.

through truss by means of four 6 x 6 inch angles bolted to the principals, as shown in Fig. 119. This construction proved very satisfactory both for strength and facility in erecting.

Fig. 121 is a view of the church taken after completion. Fig. 122 is an isometric view showing the full diagonal truss and one of the half trusses hung from it, as used by the author in another church, the span of the diagonal truss being 53 feet. In this instance the half trusses were supported by double stirrup bearing at the apex of the full truss.

CHAPTER VIII.

WIND BRACING OF BUILDINGS, TOWERS AND SPIRES.*

It is generally considered that buildings of substantial masonry construction with permanent partitions do not require any special wind bracing unless the hight exceeds one and one-half times the width of base, and for hights of from one and one-half to two times the width of base what bracing is required can generally be provided by a few brick partitions, or by substantially braced wooden partitions.

Frame buildings are not as rigid as those with masonry walls, but a well framed wooden building having cross partitions extending from first floor to top story, and not over 18 feet apart, will not require special bracing unless the hight exceeds one and one-half times the width of base. If the building has high stories and few, if any, partitions some form of bracing will be desirable, if not absolutely necessary.

For frame buildings having a hight of two or more times the width some provision should be made for wind bracing. The method of computing wind stresses can best be explained in connection with an example.

Example of Wind Stresses and Bracing.—Let Figs. 123 and 124 be the plan and cross section of a long and narrow dormitory or rooming house. The floors of such a building should be supported by posts and girders, as indicated in solid black. There

---

*The first portion of this chapter was written in response to a communication in *Carpentry and Building,* asking how the wind stresses on the sides of a building are estimated.

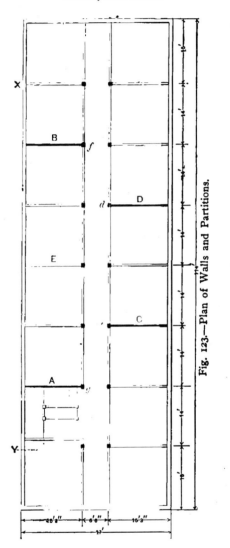

Fig. 123.—Plan of Walls and Partitions.

would, of course, be no necessity for bracing the building endways, so all we need to consider is the cross bracing of the building. We will assume that the building is exposed to the wind on both sides. The first step is to decide on the wind pressure per square foot.

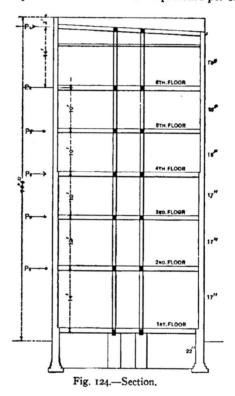

Fig. 124.—Section.

It is customary to figure wind stresses on a basis of 30 pounds per square foot of vertical surface. The actual pressure on buildings often reaches 40 pounds and sometimes 50 pounds, but it is generally assumed that the natural stiffness of the building will care for all wind stresses above 30 pounds. It is also assumed that the

wind pressure will be concentrated at the floor lines, in the same
way that the loads on a truss are concentrated at the joints.

These wind loads will be represented by the arrows P_1, P_2, P_3,
etc., in Figs. 124 and 125. The wall area contributory to P_1 will
be equal to the distance down from the top plus one-half the distance to P_2 times the length of the building; the area contributory
to P_2 equals one-half the distance from P_1 to P_3 times length of
building, and so on. The contributory area multiplied by 30
pounds will give the pressure in pounds. In this example we will
assume that the wind pressure on the end panels is resisted by the
ends of the building, and that only the six interior panels need be
figured for wind load.

P^1 will then be equal to $2' + 7' \times 84' \times 30$ pounds $= 22,680$
pounds. P_2 will be equal to $7' + 5' \times 84' \times 30$ pounds $= 30,340$
pounds.

DIFFERENT METHODS OF BRACING.

There are two methods generally adopted for interior bracing:
First, by diagonal systems, placed in partitions; and, second, by
knee braces.

Both systems require continuous posts from basement to top
story, and struts or girders in each floor, connected to the posts.
In the diagonal system the struts in the floors are connected by
diagonal rods or bars, provided with turnbuckles, as in Fig. 125.
With the knee brace system the struts are connected to the posts at
each end by knee braces, or in steel buildings by gusset plates,
which must be capable of resisting both tension and compression.

Knee bracing requires posts in the outside wall as well as interior columns.

The diagonal system is considered the most economical when it
can be employed—*i. e.*, when it will not interfere with the interior
arrangement of the building.

Diagonal Systems should be symmetrically located in relation to
the plan, so that there will be no tendency to twist. In the building in question we will locate diagonal systems in the partitions at
A, B, C and D of Fig. 123.

A diagram of one system is shown in Fig. 125. The struts S_1, S_2, S_3, etc., will be in compression, also the leeward columns, C_1, C_2, C_3, etc., while the diagonals and the windward columns will be in tension. With the wind from the left the diagonals shown by

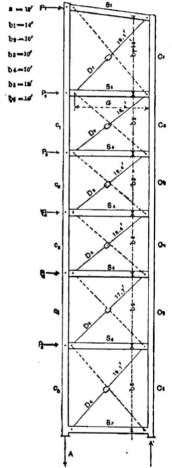

Fig. 125.—Diagram of Wind Bracing.

the full lines will be brought into action, and with wind from the right those shown by the dotted lines will be needed. As the wind may blow from either direction, both sets of diagonals will be required, and all columns must be capable of resisting tension and compression.

If the building in question were built of frame, there should be a post in the outer wall opposite each strut, and both the exterior and interior columns should be spliced so as to be continuous from bottom to top, the struts and girders being supported by steel brackets bolted to the posts. The most practical construction is to make the struts S_1, S_2, etc., of two pieces, so that the diagonals may pass between them and be attached to a pin bolt. In steel buildings a pair of channels is commonly used for the struts.

The stresses in the different members of a diagonal system are computed as follows, the dimensions being those shown on the Figs. P_1, P_2, etc., representing the entire wind load from X to Y of Fig. 123.

LOADS:

Load $P_1 = 7' + 2' \times 84' \times 30$ lbs. $= 22,680$ lbs.
Load $P_2 = 5' + 7' \times 84 \times 30$ lbs. $= 30,240$ lbs.
Load $P_3 = 10' \times 84 \times 30$ lbs. $= 25,200$ lbs.
Load $P_4 = 10' \times 84 \times 30$ lbs. $= 25,200$ lbs.
Load $P_5 = 11' \times 84 \times 30$ lbs. $= 27,720$ lbs.
Load $P_6 = 13' \times 84 \times 30$ lbs. $= 32,760$ lbs.

STRESS IN STRUTS (COMPRESSION):

Comp. in $S_1 = P_1$ $= 22,680$ lbs.
Comp. in $S_2 = P_1 + P_2$ $= 52,920$ lbs.
Comp. in $S_3 = P_1 + P_2 + P_3$ $= 78,120$ lbs.
Comp. in $S_4 = P_1 + P_2 + P_3 + P_4$ $= 103,320$ lbs.
Comp. in $S_5 = P_1 + P_2 + P_3 + P_4 + P_5$ $= 131,040$ lbs.
Comp. in $S_6 = P_1 + P_2 + P_3 + P_4 + P_5 + P_6 = 163,800$ lbs.

STRESS IN DIAGONALS (TENSION):

Tension in $D_1 =$ stress in $S_1 \times \dfrac{D_1*}{a} = 22,680 \times \dfrac{19.1}{13} = 33,340$ lbs.

Tension in $D_2 =$ stress in $S_2 \times \dfrac{D_2}{a} = 52,920 \times \dfrac{16.4}{13} = 66,680$

Tension in $D_3 =$ stress in $S_3 \times \dfrac{D_3}{a} = 78,120 \times \dfrac{16.4}{13} = 98,430$

*The length of D_1, D_2, &c., should be taken. D_1, a and b should all be in the same unit of measurement, either feet or inches.

Tension in D_1 = stress in $S_4 \times \dfrac{D_4}{a} = 103,320 \times \dfrac{16.4}{13} = 130,180$ lbs.

Tension in D_1 = stress in $S_1 \times \dfrac{D_3}{a} = 131,040 \times \dfrac{17.7}{13} = 178,210$

Tension in D_4 = stress in $S_4 \times \dfrac{D_6}{a} = 163,800 \times \dfrac{19.1}{13} = 240,790$

STRESS IN LEEWARD COLUMNS (COMPRESSION):

Comp. in C_1 = stress in $S_1 \times \dfrac{b_1}{a}$ 24,400 lbs.

Comp. in C_2 = stress in $S_2 \times \dfrac{b_2}{a}$ + stress in C_1 = 65,150 lbs.

Comp. in C_3 = stress in $S_3 \times \dfrac{b_3}{a}$ + stress in C_2 = 125,300 lbs.

Comp. in C_4 = stress in $S_4 \times \dfrac{b_4}{a}$ + stress in C_3 = 204,850 lbs.

Comp. in C_5 = stress in $S_5 \times \dfrac{b_5}{a}$ + stress in C_4 = 325,800 lbs.

Comp. in C_6 = stress in $S_6 \times \dfrac{b_6}{a}$ + stress in C_5 = 502,170 lbs.

Tension in c_1 = comp. in C_1; in c_2 = comp. in C_2, and so on.

These stresses are the TOTALS for the entire six interior panels, and if we use four sets of bracing, as at A, B, C and D, the stresses for each of the four systems will be one-fourth of those given above.

If we were to use still another set of bracing at E, then the stresses for each system would be one-fifth of those found above.

Comparing the formulas for the stresses, it will be seen that the stress in any strut is equal to the wind load at that floor plus all of the loads above, also that the column loads accumulate rapidly toward the bottom.

In proportioning the parts of a wind brace system it is customary to allow greater unit stresses than are used for ordinary live loads. Thus the diagonals are generally proportioned to a unit stress of 20,000 pounds to the square inch.

The compression in the leeward columns must be added to the usual live and dead loads, to get the size of the column.

Thus with four sets of wind bracing, the compression in one first-story column, from the wind pressure, will be 125,540 pounds. Allowing 80 pounds per square foot of floor for dead and live loads and partitions and 60 pounds for roof and upper ceiling, the load produced thereby on the first-story columns will be 66,550 pounds. Therefore the columns must be proportioned to resist 125,540 + 66,550 = 192,090 pounds. A smaller factor of safety may be used, however, than ordinarily.

With the wind from the left, the columns at d and e, Fig. 123, will be in tension, and with the wind from the right those at f and g will be in tension, and, according to our figures, the tension due to the wind load will be greater than the compression from the floors and roof. If this condition really existed in the building it would be necessary to anchor the lower columns to the foundations. In a building such as we are considering the author does not believe that this is necessary, although if the building were built of frame it would be advisable to anchor the outer columns to the foundations.

TO FIND WHETHER THERE IS DANGER OF A BUILDING BEING OVERTURNED.

There is sometimes danger of narrow and tall frame buildings being overturned by the wind, and to determine whether such danger exists, multiply the area of the entire side of the building by 30 or 40 pounds (according to the exposure), and this product by one-half the hight of the building, and divide by one-half the width of the building. If the quotient is greater than the weight of the building, there is some danger of its being overturned.

Example.—The area of one side of the building shown by Figs. 123 and 124 = 74' × 114' = 8436 square feet. Multiplying by 30 pounds per square foot, and that product by one-half of the hight, or 37 feet, we have 9,363,960 pounds as the moment tending to overturn the building. Dividing this moment by 18½ (one-half of the width) we have 506,160 pounds.

The weight of the building is equal to the weight of the walls, floors, roof and partitions.

For a frame building the walls may be figured at 20 pounds per

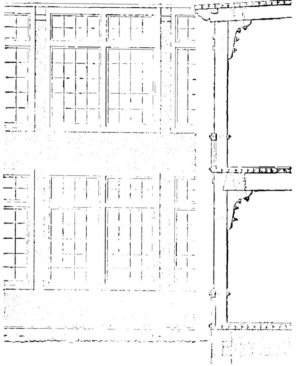

Fig. 126.—Section of Mill Building with Wood and Plaster Walls Supported by the Frame—Showing Method of Wind Bracing.

square foot, the floors, including partitions, at 40 pounds, and the roof and ceiling at 40 pounds.

In this building there are 302 x 74 feet of wall, equals 22,348 square feet, which at 20 pounds will equal 446,960 pounds.

The five floors and roof, including partitions, will weigh about $36 \times 113 \times 40 \times 6 = 976,320$ pounds, and the total weight of building, neglecting first floor, will be $446,960 + 976,320 = 1,423,280$ pounds, and as this greatly exceeds the quotient found above, there is no likelihood of the building being overturned.

Frame buildings without cross partitions should be braced by knee braces, when the hight of the building exceeds the width. Fig. 126 shows one method of bracing mill buildings.

The above covers in a general way the method of figuring wind stresses. Those readers who wish to study the subject further will find a complete discussion of the different methods of wind bracing, with formulas for the same, in Chapter XXVIII of *The Architects' and Builders' Pocket Book*.

FIGURING STRESSES IN A TRESTLE.

Problem presented by A. P. S., Williamsport, Pa.—We desire to burn shavings from the planing mill and carpenter shop, which is located 553 feet from the power house, and the idea is to convey them through two pipes placed one above the other, as shown on the accompanying blue print, Fig. 127, the top pipe being 19 inches and the bottom one 28 inches in diameter. The hight from the ground to the center of the lower pipe is 42 feet and the pipes are spaced 6 inches apart. We desire to put up just as neat and light structural steel supports as the conditions will permit, following out the design as indicated in the drawing and making a sufficient allowance for wind pressure.

I would like to have Mr. Kidder show by formulæ what strain is on each member of the truss, including the diagonals, also the distance apart the supports should be located. I would state that the pipes are made of No. 20 galvanized iron and weigh respectively 12 7-10 pounds and 8 7-10 pounds per foot. At present these pipes are supported by ugly wooden trusses, which not only take up much valuable room, but present anything but a handsome appearance. The end thrust on the pipes is taken care of by being rigidly fastened to the two respective buildings at each end of the line.

Answer.—The stresses in a trestle, as in a truss, can usually be more quickly and easily found by the graphic method, and hence I

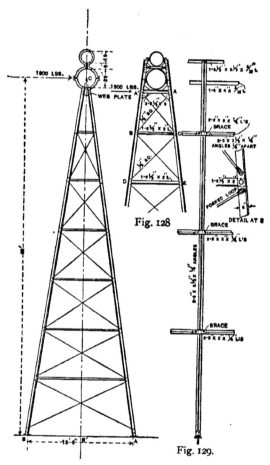

Fig. 128

Fig. 129.

Fig. 127.—Construction Proposed by "A. P. S."
Fig. 128.—As Suggested by Mr. Kidder.
Fig. 129.—Side Elevation of Trestle.

shall use this method in explaining the nature of the stresses and
how they are found.

After having determined on the outline of the trestle, the next
step is to figure the load or pressure it must be capable of resisting with safety. The trestle has to sustain two loads, a vertical
load, consisting of the pipes, sawdust and weight of the frame
itself, and a horizontal pressure due to the wind pressure against
the pipes. We will first compute the wind load and the stresses
due to it, as they are the more important.

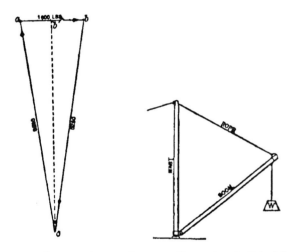

Fig. 130.—Stress Diagram.   Fig. 131.—Illustrating the Principle.

From the bottom of the large pipe to the top of the small one is
4 feet 7 inches. As the pipes are round and there is a space between them, an allowance of 20 pounds per square foot will probably be sufficient, which would make the total wind pressure about
92 pounds per running foot. For convenience in figuring and to
be on the safe side, we will call the wind load 100 pounds per
lineal foot of conduit. Economy will require that the trestles be
placed at least 16 feet apart, I should say, which will make the

wind pressure to be resisted by each trestle 1600 pounds. The frame itself is so light that we may neglect the wind pressure on it.

Now to find the stresses due to this 1600 pounds wind pressure draw a horizontal line *a b*, Fig. 130, and make its length equal to 1600 pounds at some scale, say, 800 pounds to the inch. Then from one end, *b*, draw a line parallel to the right leg of the trestle, and through the other end, *a*, a line parallel to the left leg, and continue the lines until they intersect at *c*. Then the line *b c* will represent the compression in the right leg of the trestle, and *a b* the tension in the left leg. In the actual trestle, Fig. 127, the two legs do not quite meet, but the web plate would give the same result as though they did meet.

In a triangular frame the stresses found above exist in the entire length of each leg, and theoretically *there will be no stress in any of the bracing*, and if the two legs were made of sufficient size to resist the compressive stress, considered as a long strut, no bracing of any kind would be required. Practically, by inserting horizontal braces we stiffen the legs, so that only their unsupported length need be considered in determining their size. The author knows of no way in which the stress in these horizontal braces can be computed, and he does not think that the diagonal braces would be required at all. To show that there is no stress in the bracing, we have only to consider the case of a boom derrick, Fig. 131, which is exactly the same construction as the trestle, only the load is vertical, and the axis of the triangle is horizontal. It is a matter of experience that no bracing is used between the rope and boom, but the boom is made large enough to resist the entire compressive stress without bracing.

Practically, I do not think the method of supporting the pipes shown by "A. P. S." to be a good one, because I think the wind would bend the iron work which supports them.

I would recommend that the top of the trestles be framed as in Fig. 128, and horizontal angles run from trestle to trestle to support the pipes, by means of bands. In this way no dependence is placed upon the strength of the pipes, and the trestles can easily be placed 16 feet apart.

If we consider the trestle, Fig. 128, as cut off at A', then there will be stresses developed in the bracing, and these stresses can

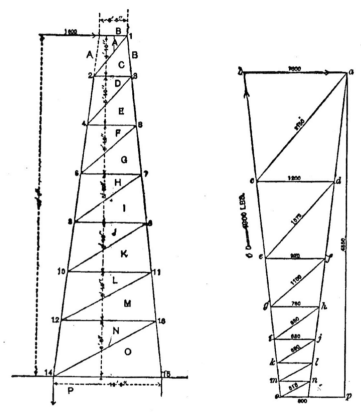

Fig. 132.—Diagram of Trestle.   Fig. 133.—Stress Diagram.

readily be found graphically. Let Fig. 132 represent the center lines of the members of the trestle with the wind from the left.

It should be noted in this connection that only one set of diagonals can act at one time. With the wind from the left the diagonals should be placed as in Fig. 132; with wind from the right an opposite set of diagonals will be required, and as the wind may blow from either direction, it is, of course, necessary to put both sets of diagonals in the structure, but for computing the stresses only one set should be shown.

To draw the stress diagram, Fig. 133, commence with the horizontal line $a\ b$, equal in length to 1600 pounds. (A scale of 1000 or 800 pounds to the inch should be used.) Then from a draw a line parallel to the diagonal A C, and from b a line parallel to B C, the two lines intersecting at c. [Note that lines in Fig. 132 are denoted by the letters at each side of them, capital letters always referring to Fig. 132 and small letters to Fig. 133.]

Then $a\ c$ denotes the stress in the upper diagonals, and $b\ c$ the stress in the top section of the right-hand leg. There is no stress in the top section of the left leg. The stress in A B will, of course, be 1600 pounds.

Next, at joint 2, we have the stress in C A, represented by the line $c\ a$, and at this joint the stress acts up. In a trestle we must take the members in order, going around to the *left* so that the next member is A D. From a draw a line parallel to A D, and from c a line parallel to C D, the two intersecting at d. At joint 3 we have the known stresses $b\ c$, $c\ d$, and draw $d\ e$ and $b\ e;\ b\ e$ lies over $b\ c$, but they should be considered as two different lines. The stresses at joint 4 are $e\ d$, $d\ a$, $a\ f$ and $f\ e$. In the same way the stresses are found for the remaining joints. At joint 14 we have the stress $o\ n$, $n\ a$, a stress $a\ p$, representing the anchorage required to keep the trestle from overturning, and the horizontal line $p\ o$, which represents the horizontal force tending to blow the trestle sidewise. This force, it will be noticed, is just half the wind pressure. Now by scaling the lines in Fig. 133 we obtain the amount of each stress in pounds, as indicated by the figures. The right leg and all of the horizontal braces are in compression. The left leg and the diagonals are in tension.

To the compression in the leg due to the wind pressure should be added the compression due to the dead load. The dead load I estimate as follows:

	Pounds.	Pounds.
Weight of 16 feet of upper pipe, at 8.7 pounds	139	
16 feet of sawdust, at 12 pounds	192	
Two 3½ x 3½ x 5-16 inch angles, 32 feet, at 7.1 pounds	227	
16 feet of large pipe, at 12.7 pounds	203	
Sawdust in same, at 25 pounds per lineal foot	400	
Two 4 x 4 x 5-16 inch angles, 32 feet, at 8.2 pounds	262	
		1,423
Weight of frame:		
Four 3 x 2½ x ¼ inch angles, 45 feet long, 180 x 4.5	810	
12½ feet of 3 x 2½ inch angles, at 4.5	56	
27½ feet of 2½ x 2 inch angles, at 3.7	102	
22 feet of 3 x 2½ inch angles, at 4.5	99	
134 feet of ½ x ½ inch bars, at 0.85	114	
		1,181
Total dead load		2,604

The compression in each leg will be one-half of this increased in the proportion that the length of the leg bears to the vertical hight. Without stopping to compute it, we will call the stress in each leg due to vertical load 1400 pounds, and as the wind stress in the bottom section of the leg is 4900 pounds, each leg must be computed to sustain 6300 pounds in the bottom section. As this will require only very light angles, it will be best to make the legs of the same size their entire length. The dimensions given on Figs. 2 and 3 are as light as the author would recommend.

Each diagonal should be provided with a turnbuckle. The weight of shavings was arrived at by considering the pipes filled, and by actual trial I find the weight of loose mill shavings to be about 6 pounds per cubic foot.

WIND BRACING OF FRAMED SPIRES.

A tall spire in a gale of wind is very much like a cantilever set upright and the principles involved in its construction are the same as those upon which trusses are designed, except that with a spire the destructive force acts horizontally instead of vertically.

A framed spire may fail in either one of two ways:

Fig. 134 represents a spire which is strong enough in itself, but which on account of its not being anchored to its foundation is being overturned and pushed off the tower.

Fig. 135 represents a spire which is sufficiently anchored but insufficiently braced.

Fig. 136 represents the tendency of the wind pressure on the spire to rack the tower. In this case failure occurs in the tower and not in the spire.

Of course, a tower and spire may be weak in all three particulars, but actual failure will usually take place in one of the ways above illustrated.

Resistance to Overturning.—In the case of a spire whose hight is not greater than twice the width of its base, the weight of the

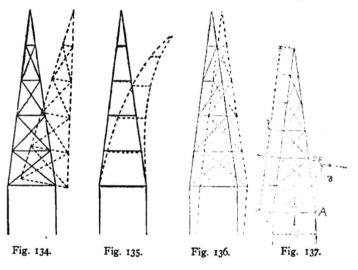

Fig. 134.   Fig. 135.   Fig. 136.   Fig. 137.

frame work and covering will generally be sufficient to prevent its overturning, although all spires should be bolted to the walls to some extent for the purpose of strengthening the wall and overcoming the tendency to slide.

When the hight of a spire exceeds two and one-half times the width of its base, it will generally be necessary to depend upon rods or bolts carried well down in the masonry to prevent the windward side being lifted from its foundation.

The method of determining the size and length of these bolts is as follows:

Let Fig. 137 represent the framework of a square spire 16 ft. wide at the base and 60 ft. high, measured on the rafter. Assume that the wind is from the left. Now it can be readily seen that the tendency of the wind, blowing against the left-hand side of the tower, is to cause the framework to revolve about the lower point A. This tendency is in part resisted by the weight of the framework and roof covering, which act through the center of gravity of the spire.

The force tending to overturn the spire equals the entire windward·side of the spire, multiplied by the pressure per square foot.

In calculating the wind pressure on roofs it is customary to assume 40 pounds as the maximum horizontal pressure per square foot.

As spires, however, are usually exposed to the full force of the wind, the author recommends that 50 pounds be used as the normal wind pressure on spires having a pitch exceeding 60 degrees.

The area of one side of the spire shown in Fig. 137 = 480 square feet. Multiplying this by 50 we have 24,000 pounds as the total wind pressure on the spire.

The resultant of this pressure acts through the center of gravity of the side, which in a triangle is at one-third the hight, measured from the base, or in this case 20 ft. As wind pressure is always considered to act at right angles to the surface against which it blows, the resultant will act in the direction of the line R F, and will have a moment equal to the whole wind pressure multiplied by the arm X. The distance X in the figure measures 17.8 feet. The moment tending to overturn the spire will therefore be 24,000 × 17.8 = 427,200 foot pounds.

The weight of the framework, if of wood, will be about 10 pounds per square foot of surface, and the sheathing and shingles will weigh about 5 pounds more, giving a total of 15 pounds per superficial foot for a wooden spire covered with shingles. If covered with slate the weight may.be taken at 18 pounds.

Assuming that the spire under consideration is covered with

180

STRENGTH OF BEAMS,

shingles, we will have for the total weight $4 \times 480 \times 15 = 28,800$ pounds. Its resultant will act through the center of the base, and will therefore have a moment about the point $A = 28,800 \times 8 = 230,400$ foot pounds. This moment acts in the opposite direction from that produced by the wind, and hence should be subtracted from it: $427,200 - 232,400 = 196,800$ foot pounds.

The difference between these two moments represents the force which must be resisted by the anchors on the windward side, as at S. These rods will act with an arm equal to the width of the base, in this case 16 ft., and we should therefore divide 196,800 by 16 to get the stress on the rods at S. Performing the division we have 12,300 pounds as the upward pull on the rods. If two rods are used we will have 6,150 pounds stress on each rod, which will require a diameter of 1 inch if the rod is not upset.

To hold the rods there must be a weight of masonry on the windward side = 12,300 pounds, or about 110 cubic feet. If the wall is 16 inches thick the rods must be imbedded at least 5 feet.

By studying Fig. 137, it will be seen that the wind moment increases with the relative hight of the spire much faster than the weight moment, as the arm of the latter remains constant for a given width of the base, while the former increases with the hight, due to the raising of the center of gravity of the side and the diminishing of the angle of the resultant with a horizontal line.

Thus, if the length of the rafter in the above example were reduced to 39 ft., keeping the same width of base, we would have 150,540 for the wind moment and 149,760 for the weight moment. And by taking off 1 foot more the spire would be just about in equilibrium. As the wind often acts with considerable impact, however, there should always be some excess of safety.

It will also be seen from the above that the stability is increased by adding to the weight of the tower. Thus, in the case above, where the rafters were 39 ft. long, if the roof were covered with slate, the additional weight would increase the weight moment to 179,712, or considerably in excess of the wind moment.

In an old English spire, which swayed more than was thought

safe, a heavy timber was suspended inside from the top, with apparently beneficial effect.

On an octagonal tower the wind pressure is assumed to act against a surface equal to the vertical section taken through the center of the spire, so that the uplift on the windward side is the same as for a square spire.

Effect on the Tower.—As shown by Fig. 136, the effect produced on the tower by the wind pressure on the spire is a tendency to rack the tower sideways; this tendency is also increased by the wind blowing against the tower itself. As a rule, towers which support a spire have at least two sides built into the building of which the tower forms a part, so that the leeward side of the tower is well braced, at least to the hight of the main walls. If the tower is of frame construction the walls should be braced in a similar manner to that hereafter described for the spire, at least down to the hight of the main walls, and all the uprights should be spliced so as to resist tension, that the ties connecting the spire frame with the walls may draw on the entire weight of the tower.

In brick or stone towers the bonding of the side walls is generally depended upon to resist the tendency to rack. Whenever the hight of a spire exceeds 40 ft. the upper portion of the tower walls should be at least 16 ins. thick, unless the corners are strengthened by heavy buttresses.

Another tendency resulting from the wind pressure on the spire is to lift the top of the tower on the windward side, and in a masonry tower to break and crack the masonry near the top. Very often the top of the tower is so slight, owing to the architectural treatment, that it cannot be depended upon to resist the strains from the spire, and braces must be carried down below the upper openings. Examples of such construction are shown in Figs. 138 and 139. Sometimes a square framework is built inside the walls of the tower, with the lower timbers extended into the walls and the spire supported from this framework.

Bracing the Spire.—The strains in a spire may be accurately determined in the same manner as those in a trestle (see pages 175 and 176), and the size of the braces proportioned accordingly.

Generally, however, the braces are made larger and more are used than would be determined by theoretical analysis, practical judgment being as important as theoretical analysis, but it is not always safe to depend upon the former alone. When there is a lack of theoretical knowledge, material is generally used unwisely, too much being often used in some places and not enough in others. The correct method of bracing a spire is shown by Figs. 138 and 139.

Examples of Spire Framing.—Fig. 138 shows the framework of a spire designed by the author some fifteen years ago. This spire is 111 ft. 6 ins. high above the plate, and the latter is 69 ft. above the sidewalk. The total hight from sidewalk to top of finial is 190 ft. The tower is of stone, 19 ft. square, with buttresses as shown. The spire is a true octagon in section, and each of the eight sides is braced in the same way, with the exception of the lower panel, in which the bracing is omitted on four sides back of the dormers. Besides the bracing shown in Fig. 138, the spire was braced across horizontally at each purlin to prevent distortion in the octagon. At the top the eight hips are cut against a 10-inch octagon pole and bolted to it in pairs. This pole is 32 ft. long and is secured at the bottom by bolting to 4 x 6 cross pieces which are securely spiked to the hips. In the center of this pole is a 1½-inch iron rod, which forms the center of the wrought-iron finial.

The lower end of each hip is secured to the masonry by 1½-inch bolts, 6 ft. long. The plate extends the full length of each side of the tower and is bolted together and to the walls at the corners. A short piece of 6 x 6 timber is placed on top of the plate, across the corners, to receive the rafters on the corner sides of the octagon. The braces and purlins are set in 4 ins. from the outer face of the hips to allow for placing 2 x 4 jack rafters outside of them. These rafters are not shown in the figure; they were placed up and down 16 inches on centers, and spiked to the purlins and braces.

As may be seen from the engraving, the top of the tower is rather light for supporting such a high framework, and is more-

over weakened by large openings in each side. It was therefore
determined to transfer the thrust due to the wind pressure on the
spire to the corner of the tower at a point just below the sill of the
large openings. The manner in which this was done is shown by
Fig. 139, which is a diagonal section through the top of tower.
The purlins C, C, Fig. 138, were made 6 x 10 ins., set on edge and
securely bolted to the hips. From the center of these purlins on
each of the four corner sides 6 x 10-inch posts were carried down
into the tower, as shown in Fig. 139. These posts were secured
at the bottom to 10 x 10-inch timbers which were placed across
the tower diagonally and solidly built into the corners. The bracing shown was used merely to prevent the posts from buckling.
Only one pair of posts is shown in the figure. The effect of these
posts is to transmit the entire wind pressure on the leeward side
of the tower from the purlins C, C to the corners of the tower at
the bottom of the posts. The tension on the windward side is resisted by the hip rafters and the bolts by which they are anchored
to the wall. This spire has withstood many severe wind storms,
and no cracks have as yet appeared in the tower, although the
1½-inch rod in the wrought-iron finial was slightly bent during a
severe gale.

Fig. 140 shows the framing of a smaller spire, designed by the
author, in which a similar method was used for transferring the
thrust to a point about 8 ft. below the plate. In this case the spire
was square, and the posts were bolted to the inside of the hip
rafters. At the bottom the posts rest on short pieces of 8 x 8
timbers, T, T, built into the corners as shown.

In this spire the braces were run diagonally across the tower
from hip to hip, only one pair of hips with its braces being shown.

The best method of bracing such a spire would be by means of
rods placed in the sides just under the jack rafters, as shown by
the dotted lines in the side elevation. By properly tightening these
rods the spire can be readily straightened and made very stiff. In
country places, however, rods cost considerably more than timber,
so that it is generally customary to use plank for the braces.

In framing the hip rafters it must not be forgotten that one pair will always be in tension when the wind blows hard, and the splices must be made accordingly. It is also always desirable to use a center pole for the hips to cut against. This pole should

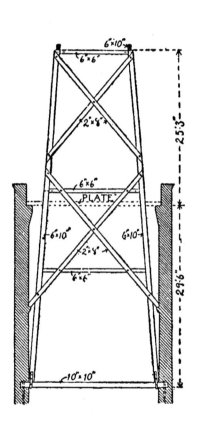

Fig. 138.   Fig. 139.

extend some distance below the peak and be stayed at the bottom
by cross pieces.

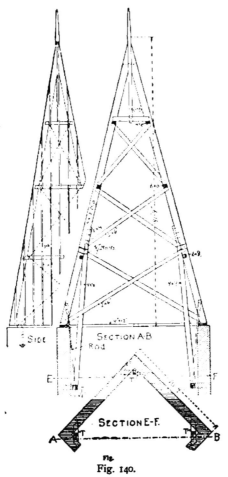

Fig. 140.

The author has found that it is generally difficult to keep the
anchor bolts for the hip rafters in the desired position, and that

instead of passing the bolts through the toe of the hips it is better
to flatten out their upper end and bolt to the hip as shown in Fig.
141. The plate should also be bolted to the wall by shorter bolts,
for the purpose of strengthening the walls of the tower and to
resist the sliding tendency.

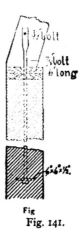

Fig. 141.

Fig. 140 shows another method of anchoring the hips. This
method has the advantage that the long bolts can be put in after
the hips are set in place and the nuts can be screwed up tight after
the bolts are fastened to the hips. When this method of anchoring
is used the plate should be secured to the wall by ¾ or ⅞-inch bolts
about 3 ft. long and at least four bolts in each side of a 16-ft.
square tower.

CHAPTER IX.

EXAMPLES OF TRUSS CONSTRUCTION, WITH CRITICISM BY THE AUTHOR.*

TRUSS FOR ROOF OF THIRTY-FOUR FEET SPAN.

Question.—Would a roof built according to Fig. 142, with trusses placed 2 feet on centers, be sufficient for a span of 34 feet? Is the timber large enough? What change would be necessary, if any? I want the hight as shown in the sketch.

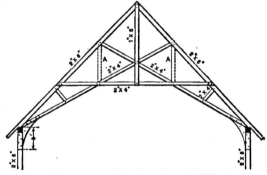

Fig. 142.—Sketch Submitted by "G. M. E." to Which the Dotted Lines Have Been Added.

Answer.—The truss shown in Fig. 142 is a good and economical type for church roofs, ranging from 30 to 36 feet in plan, *provided a tie is inserted* at A A, as indicated by the dotted lines. Without this tie a heavy cross strain is brought both on the tie and

*Most of the material in this chapter is taken from the correspondence Columns of *Carpentry and Building*, the questions being those submitted by the readers, and the answers as furnished at the time by Mr. Kidder.

rafter. The truss as shown in the original form cannot be considered safe for a span of 34 feet with a spacing of 2 feet, although it might stand for some time. By placing a 1 x 6 tie at A A and increasing the rafter to 2½ x 6, it will be safe for a shingle roof and plastered ceiling, provided, however, that enough spikes are used in the joints. What the author would recommend, however, is that a truss of the dimensions shown in Fig. 143 be employed and the

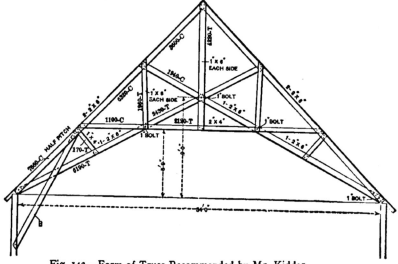

Fig. 143.—Form of Truss Recommended by Mr. Kidder.

truss spaced 32 inches on centers, the ceiling beams being strapped with 1 x 2½-inch stock, 12 or 16 inches on centers for lathing. The advantage of the truss, Fig. 143, over the truss shown in Fig. 142, is that in the former the rafters are double and the ends of the tie beams and collar beam are spiked between them. This secures a symmetrical strain without a tendency to twist. It is also better to use two boards for the vertical ties, one on each side of the truss. Where the ties intersect there should be at least one bolt reinforced by spikes. When spikes alone are used they are sometimes partly drawn by the twisting or warping of the planks.

There is also another point which should be considered in placing such a roof over a large room, and that is the danger of the building collapsing from a heavy wind pressure against the side of the structure. If the room is more than 30 feet long without any break or buttresses to stiffen the sides, or cross partitions, the trusses should be braced from the walls, as indicated by the brace B in Fig. 143. The author has in mind a brick veneered frame church, 80 feet long by 42 feet wide, roofed with trusses somewhat similar to Fig. 143, which was blown down before it was entirely completed, owing to a lack of bracing of the walls.

The relative stresses in the different pieces of this truss, Fig. 143, are indicated in pounds in connection with each piece, calculated for a spacing of 32 inches and allowing 30 pounds pressure per square foot for wind and snow, the ceiling being plastered. The letter "T" denotes that the piece is in tension, while "C" denotes compression. These figures show which parts of the truss are called upon to bear the most strain. The tie beams and rafters have to be computed for a transverse strain and also for a direct stress.

POOR CONSTRUCTION OF CHURCH ROOF.

From C. E. B., *Judson, Ind.*—I inclose a rough drawing of a church roof that I erected last summer, and, as I never built one before of this style, I contended with the Building Committee that it would sag and that the walls were too small. The results have shown that my contention was correct. The roof timbers are of 2 x 5 beech, and the walls, or outside studs, 2 x 4 beech, 14 feet long. The roof was constructed as shown in the sketch, the drawings of the building being prepared by a man who was supposed to know his business. The rafters were set 2 feet apart and plastered on, with circle boards in the center and at the wall plate. What should be the proper dimensions of timbers of pine? and kindly comment upon the general construction for the benefit of others, as well as myself.

Answer.—The great defect in this roof truss is that two ABSO-

LUTELY NECESSARY members are omitted. If the ties X and Y had been put in the truss it would probably have stood up without sagging, but omitting these members throws the whole stress on the rafters and causes them to bend, as shown by the dotted lines. If the roof is still standing when this issue of the paper reaches the correspondent it is liable to collapse at any time unless the walls have been braced. Putting in the members X and Y gives the truss

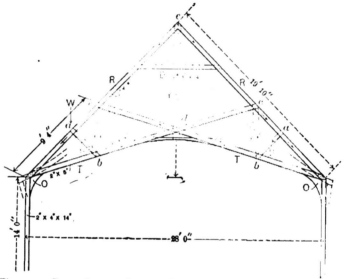

Fig. 144.—Poor Construction of Church Roof.—Dotted Lines Show Changes Suggested by Mr. Kidder.

the shape shown in Fig. 110. The necessity for ties X and Y may be more clearly understood by following the strains produced by a load at a, which are the same as those actually produced by the weight of the roof. A load at a produces a thrust in the rafter and brace B. The piece X takes the thrust from b and carries it to c. At c part of the pull in X is taken by the rafter and part by the brace $c\ d$, as indicated by the arrows. The piece Y holds up the ties T at their intersection and transmits all of the stress at d to the apex, where it is supported by the rafters. In this way the rafters are made to carry the entire weight on the truss, and the ties T, T and Y prevent the rafters from spreading. If any one of the members with an arrow marked on it is omitted the principle of the truss is broken and a bending strain is produced, either on the rafter or tie beam, or if both X and Y are omitted a bending strain is produced on both the rafter and tie beam. The piece D does not transmit any stress, but is useful in strengthening the rafters.

As to the correct dimensions for pine, I would recommend that the rafters be made of two 2 x 6 planks, with the ties T and braces B spiked between. One 2 x 6 would do for T, T. For X, I would recommend two 1 x 6 boards; for Y, two 1 x 8 boards, and for D, a 2 x 4. There should also be one ¾-inch bolt in each of the joints o and d, and as many spikes as practicable. The boards forming the tie Y should be spiked to the rafters as strongly as possible.

IS THE ROOF TRUSS SAFE?

From WARE, *Massachusetts.*—I am an interested reader of the paper and am desirous of obtaining information as to the safety of the roof truss, shown by Fig. 145. Referring to the drawing, I would say that the building is 42 feet span and the truss is full width, for the partition of 7 feet 8 inches is not supposed to carry the weight of the truss or any part of it. The owner of the building which is to be erected does not want the ceiling of the hall cut around the room at the top any more than shows at the foot of the rafter A. Again, he claims that the truss takes too much timber and he wants to take off one of the 2 x 6 inch rafters where two 2 x 6's are shown. Where one 2 x 6 is shown he wants to substitute a 2 x 4 and use a 2 x 6 in place of the 2 x 10. As a consequence we cannot agree. I am inclined to think there is a weak place in the truss at A, for when the partition on the opposite side is set up, as shown, it will have to take some of the weight of the truss and then the load of the truss will not be equally distributed and will tend to break the rafter, I think, at A. The trusses are

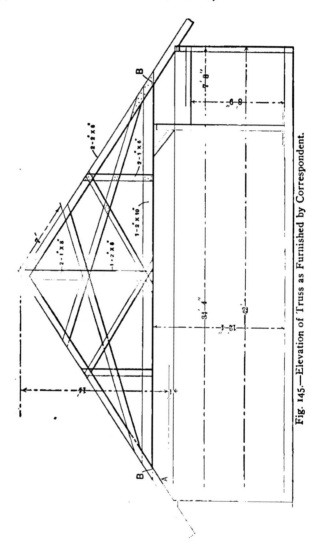

Fig. 145.—Elevation of Truss as Furnished by Correspondent.

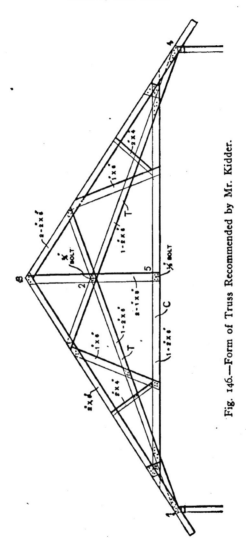

Fig. 146.—Form of Truss Recommended by Mr. Kidder.

placed 24 inches on centers and are intended to carry a shingle roof with lath and plaster ceiling, the ceiling being furred.

Answer.—A truss built as shown in Fig. 145 would be absolutely unsafe and would probably fall at the first heavy snow storm, if not before. The greatest defect is that the whole weight of the roof is brought as a cross strain on the rafters at the points B B of Fig. 145.

The truss should be constructed as shown in Fig. 146, the dimensions thereon given being the very least consistent with safety, and plenty of spikes should be used in the joints, while bolts should be placed in the joints marked 2 and 5. The main ties T T should be placed between the rafters. The piece C should be nailed to the side of the ties and cut against one of the rafters and a board nailed over the joint. The pieces T T must be in one length. A 2-inch block should be put between the rafters at the apex and the 1 x 8's spiked with at least eight 40d spikes on each side. In this truss the greatest strains come at the joints 1, 2, 3 and 4.

TRUSS FOR HALL WITH ELLIPTIC CEILING.

From W. S., *Walcott, Iowa.*—I send herewith a drawing (Fig. 147) showing a truss suitable for halls, etc., with an elliptic ceiling, and in this connection would state that I have built two halls, both of the same size, 46 x 64 feet, with a stage at the end. I have one of these trusses every 8 feet, cross braced with 2 x 8's on the straining beam to keep the building from swinging in the middle by heavy winds. The pieces A A A, etc., are 2 x 6's, running from one truss to the other, and are suspended by 3 x 4's, bolted on the truss and placed 4 feet on centers. The studdings are notched, ⅜-inch, at X and X', to receive the ribs, which are of ⅞ x 3 inch stuff. These are bent and nailed on the 2 x 6's marked A A, etc., and placed 16 inches on centers to receive lath, making a very light, yet strong ceiling. The purlins are 6 x 8, bolted under the principal rafter. There are three 2 x 6 rafters placed between every truss 2 feet on centers.

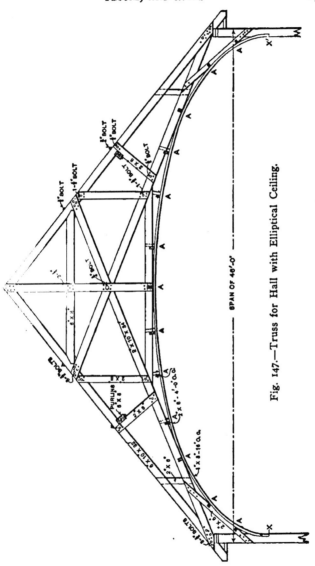

Fig. 147.—Truss for Hall with Elliptical Ceiling.

These trusses have given perfect satisfaction thus far, and I send the sketch with descriptive particulars in the hope that the matter may possibly be of interest to some of the readers of the paper. I would like very much to have Mr. Kidder say a few words in the way of criticism regarding the construction of the truss. The pitch of the roof is 10 inches to 1 foot, is covered with ⅜-inch sheathing and 5 to 2 red cedar shingles. The drawing so

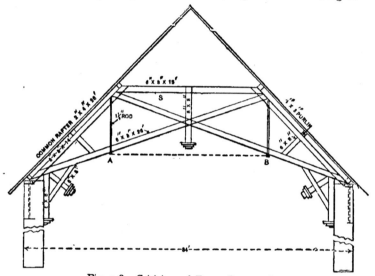

Fig. 148.—Criticism of Truss Construction.

clearly shows the construction employed that further explanation would seem to be unnecessary.

Comment.—This truss is well designed, and is perhaps the most economical construction for an elliptic ceiling.

CRITICISM OF TRUSS CONSTRUCTION.

From FRANK E. KIDDER, *Denver, Col.*—Referring to the truss submitted by "C. C. J." (Fig. 148), I would say that the form of truss shown, while a true truss, *is not a desirable one,* owing to the

enormous strain developed in the straining beam S. For the span
and spacing given, taken in connection with a shingle roof, I think
the truss will stand up, but the straining beam should be braced
laterally by a beam running from truss to truss. The only duty of
the center post is to brace the straining beam, in order to prevent
its buckling up or down, and this should be made of two planks,
bolted to the straining beam and tie beams. The straining beam
should not be cut at its center.

The strength of the truss will be very greatly increased by placing a rod from A to B, as indicated by the dotted line, and there
will then be no danger of the truss settling to any appreciable extent, or if the rod is considered objectionable, from the standpoint
of appearance, the main rafters should be extended to meet at the
peak, and the center post or tie planks also extended to the peak,
thus forming a true scissors truss, similar to that shown in Fig.
110. The straining beams could then be omitted.

ROOF TRUSS FOR A HALL.

From J. B. P., *Hawkeye, Iowa.*—I inclose a sketch (Fig. 149)
of a rafter truss which we used with satisfactory results on an
I. O. O. F. hall at Randalid, Iowa, that was 24 x 45 feet in size,
with a 13-foot ceiling in the center and about 9 feet at the side
walls. The building covered an area of 24 x 70 feet and had 22
foot posts, with storeroom below the hall. We wired the plates
together before we commenced to raise the rafters, and, by the
way, we put them together over a pattern and raised them in one
piece, making the center one a little narrow, so that they drew on
the plate. I do not believe the frame moved $\frac{1}{8}$ inch after we
loosened the wires.

FORM OF TRUSS FOR 100-FOOT BUILDING.

From T. P. P., *Housatonic, Mass.*—Will some reader be kind
enough to give me a little information regarding the form of truss
to be employed in order to sustain a frame building which extends

over a river? The structure is 100 feet long, 20 feet wide and 18 feet high, consisting of two stories. The first story is about 8 feet in the clear and the second story about 10 feet. The structure is supported in the middle, thus making two spans of 50 feet each. The timbers are to be 8 feet on centers, and the floor and roof planks to be 16 feet long and 3 inches thick. The sides are to be planked 2 inches thick, the boards running up and down and covered with clapboards. The floor planks are to be covered with maple ⅞ of an inch thick. The roof is about 3-inch pitch and the weight, besides the building, will be about 120,000 pounds.

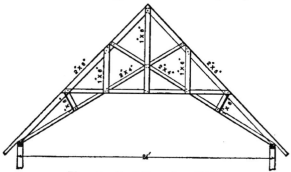

Fig. 149.—Roof Truss for a Hall.

Answer.—A building such as that described by "T. P. P." is, in effect, a double deck bridge covered by a roof and enclosed at the ends. The best form of wooden truss for supporting such a building is unquestionably that shown in Fig. 150, representing one-half of a side elevation of the building. For the construction of the floors, roof and walls described by the correspondent, the method of framing indicated seems best adapted for the purpose. Fig. 151 represents a half plan of the roof; Fig. 152 is a cross section through the building; Fig. 153 is a half plan of the lower floor clearly indicating the tie rods, while in Fig. 154 is a plan of the tie beam, made of five 2 x 10 inch plank the full thickness. All truss timbers are of Oregon fir or Southern yellow pine.

The trusses have been computed for a load on each floor of 60

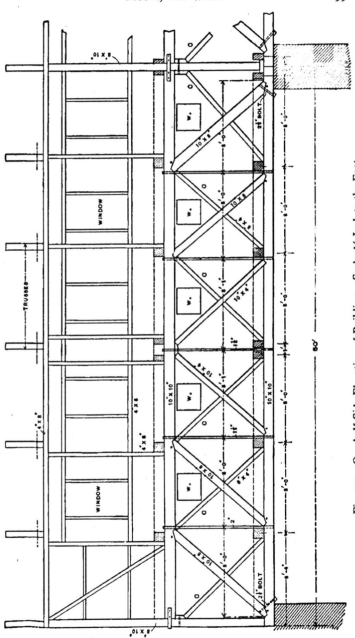

Fig. 150.—One-half Side Elevation of Building.—Scale, ⅛ Inch to the Foot

pounds per square foot, exclusive of the weight of the floor itself, and for a total roof load of 30 pounds per square foot. A load of 60 pounds per square foot is equivalent to 230,000 pounds for both floors. This is nearly twice the load given by the correspondent, but it does not seem safe to design a building of this character for a load of less than 60 pounds per square foot, and for many purposes even this would be insufficient.

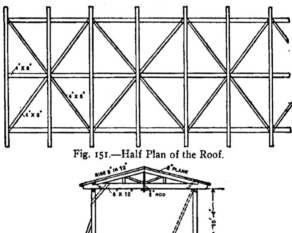

Fig. 151.—Half Plan of the Roof.

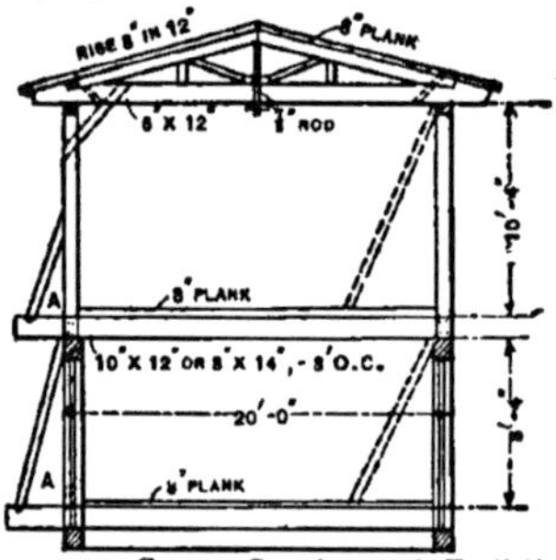

Fig. 152.—Cross Section of Building.

In order to support this floor load the floor beams should be either 10 x 12 or 8 x 14 inches, and of either Oregon fir or long leaf yellow pine. The most severe test on the structure will proba- bly be produced by the wind, which always blows strong on a river. To prevent the building from being bent horizontally by the wind the lower floor should be trussed by diagonal rods, as indicated on the floor plan, Fig. 153. These rods should pass

through the tie beam of the trusses and be held in the middle by a
strap or hook from the floor beams. A system of wooden braces
should also be placed under the roof beams, as indicated on the
roof plan, Fig. 151. Even with this bracing the upper portion
might be blown down by "racking," and to prevent this there
should be some way of bracing the walls either by extending the
floor beams, so as to get braces on the outside, as shown in the
cross section, Fig. 152, at A-A, or by placing braces on the inside,
as indicated by the dotted lines opposite every other rod and in
both stories.

The ends of the building should have a girt at the level of the
upper floor beams, and braces from each corner post toward the

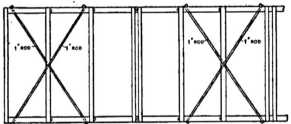

Fig. 153.—Half Plan of Lower Floor, Showing Truss Rods.

center. The second-story posts should also be well spiked to the
floor beams, the object in making the posts 8 inches wide being to
give a better chance for the spiking. It also seems desirable to
spike 2 x 6 braces to the side of the roof beams and to the posts
below, in order to prevent the roof from being lifted off the plate.

With these precautions the building should be strong enough to
stand a severe gale. There is one suggestion which should be
made in the way of a caution in regard to the trusses, and that is
about putting any transverse load on the top chords. As the
second floor is now framed the beams go over the ends of the
braces so that the top chord has to resist only a compressive stress.
If the second floor was framed by 2 x 12 inch joists placed 16
inches on centers, it would be necessary to increase the size of the
chord to 10 x 12 inches. If the tie beams are built of plank they

202 STRENGTH OF BEAMS,

should be jointed as indicated in Fig. 154, and all of the bolts specified put in. At the same time the planks should be full 2 inches thick. The center plank should be jointed at the rods. The braces marked C in the side elevation, Fig. 150, should be cut against the main braces and not halved into them.

CRITICISM OF QUEEN ROD TRUSS.

The design of a truss submitted by the correspondent (Fig. 155) is all right if the sizes of the members are sufficient to resist the loads. As the load to be supported depends upon the spacing of the trusses, or distance between centers; the kind of roofing; purpose for which the floor in the center is to be used and whether or not there is any plastering, none of which data is given, it is impossible to say whether or not the truss is sufficiently strong for the purpose. The weakest part of the truss is probably the tie beam, as that is to act both as a girder and as the main tie of the truss. If the floor is very heavily loaded the tie beam will probably be strained beyond the safe limit. It will also require good-sized bolts to hold the foot of the rafter.

Instead of doubling the rafter it would be better construction to omit the 6 x 10 and place purlins at A and B, as indicated by the dotted lines in the sketch at these points, the purlins extending from truss to truss, and on these placing 2 x 6-inch rafters running in the same direction as the 6 x 10-inch pieces. The illustration shows the truss resting on a 16-inch wall without reinforcement. I do not consider a 16-inch wall strong enough to support such a truss unless reinforced by 8 x 28-inch pilasters either inside or out.

In order to frame the truss it should be laid out full size on a floor with the tie beam crowned, say, 2 inches and the braces cut to fit. The additional inch called for by the correspondent should be obtained by tightening up on the main rods.

WHICH FORM OF TRUSS IS THE STRONGER?

From S. O. Y., *Richmond, Ind.*—I inclose sketches of two roof trusses intended for a factory building with a span of 100 feet and supported by an iron column in the center. I would like very

Fig. 154.—Plan of Tie Beam Made of Five 2 x 10 Inch Plank.

much to have Mr. Kidder criticise the construction shown, stating
which he considers the better truss for the purpose. What I
especially want to know is whether the members shown are of
proper size and arrangement to safely sustain the loads. There is
a difference of opinion between certain people and myself regarding the matter, and we have decided to accept Mr. Kidder's opinion as binding and conclusive.

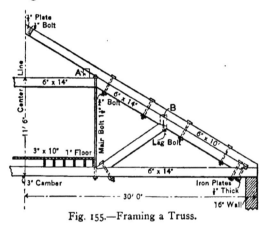

Fig. 155.—Framing a Truss.

Answer.—We have prepared engravings from the two sketches
of trusses furnished by our correspondent, and have designated
them Figs. 156 and 157 for the sake of convenient discussion.
Mr. Kidder's comments upon the trusses shown are as follows:

The arrangement of trusses indicated in Fig. 157 is the correct
one, when there is to be a support at the center. Except for the
size of one of the rods, the truss would be safe for a slate roof,
and a load of 100 pounds per square foot òn the tie beam, provided
the trusses are not more than 12 feet from center to center. The
truss shown in Fig. 156 is the correct shape for spanning the full
width of the building, although it is not strong enough for the
loads. The column in Fig. 156 would probably do more harm
than good, for if the truss bore upon it at all it would bring a
bending moment on the tie beam at X and Y, which is undesirable.

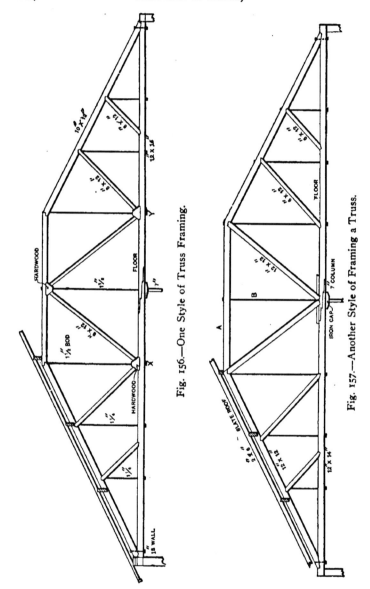

Fig. 156.—One Style of Truss Framing.

Fig. 157.—Another Style of Framing a Truss.

If, therefore, the column is to be used, there should be two trusses, as in Fig. 157. There should never be a support under a truss between the ends.

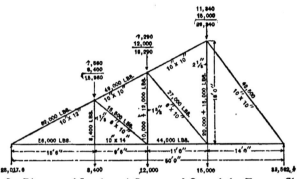

Fig. 158.—Diagram of Loads and Stresses of One of the Trusses Shown in Fig. 157.

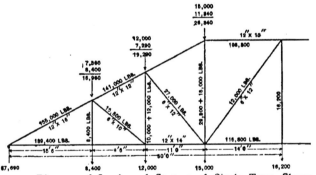

Fig. 159.—Diagram of Loads and Stresses of Single Truss Shown in Fig. 156.

The correspondent does not give the distance the trusses are to be apart. Assuming the distance as 12 feet, center to center, and that the floor supported by the tie beam will be loaded to 100 pounds per square foot (including its own weight), and allowing 19 pounds per square foot of roof surface for the weight of truss

and roof, and 26 pounds per square foot for wind and snow, the
loads and stresses on one of the trusses shown in Fig. 157 and the
single truss shown in Fig. 159 would be as indicated in Figs. 158
and 159, respectively. These latter figures also give the size of
the rods and members required to sustain the stresses. No center
rod will be required at B, Fig. 157, and the beam A should be
fitted between the trusses after they are in place.

The possible load on the column may be 79,000 pounds. The
column should be safe for this load, if the length does not exceed
16 feet. If the length is greater than 16 feet the diameter should

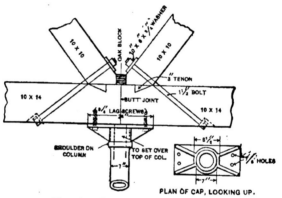

Fig. 160.—Detail of Column Support.

be 8 inches, or it would be better to use a 10 x 12 inch or 12 x 12
inch Georgia pine post.

The dimensions given for the rafter and tie beam, Fig. 159, are
for Georgia or Oregon pine, the braces to be of white pine. The
dimensions in Fig. 158 are for white pine. If the distance between trusses is greater than 12 feet the size of timbers and rods
should be increased in proportion.

The joint over the column, Fig. 157, should be made as in Fig.
160, and the joint at the wall should have three 2¼-inch bolts with
cast washers on underside of tie beam, as in Fig. 84 of Chapter
VII.

It would be good practice for those readers who have followed
the articles on "Stresses in Roof Trusses" to draw the stress

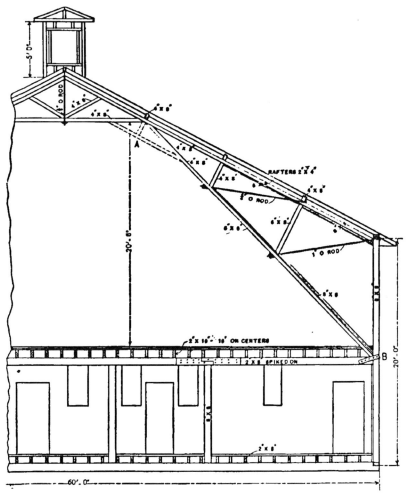

Fig. 161.—Truss for Dairy Barn.—Submitted by "W. F. C.," Elk River, Minn.

diagrams for these trusses and see how near the stresses scale to the values given in Figs. 158 and 159.

IS THE TRUSS CONSTRUCTION SAFE?

From S. C. P., *Caribou, Maine.*—I need a truss over a hall with as little elevation as possible and perfectly safe. The sketch, Fig. 162, which I send will help to make my meaning clear. My idea is to bore through the truss rafters and posts for the 1½-inch rods, the upper chord being built up with a space between the planks for the rods. I could bore through the upper chord and have it solid if that would be the better plan. Under the washers for the ¾-inch rods I would use a 2-inch plank across the girders and plate for the eye bolts would be 4 x 9 inches. The span of the truss is 45 feet and the rise is 4 feet. The top girders are made of three pieces of 2 x 14 inch spruce bolted together with 20 bolts, ⅝-inch, with cast washers. There are two rods for each truss 1½ inches in diameter. The bottom girder is 4 x 9 inches solid, the inside truss rafters 4 x 9 inches and the wrought-iron washers on the nuts are 4 x 9 inches. The estimated weight the truss will support is 10,530 pounds for roof and ceiling and 19,975 pounds with snow distributed over the surface, 13 x 45 feet. In order to make it plainer I would state we have a roof and ceiling 45 x 52 feet, with three trusses to hold the same, with the outside walls solid. We often have snowfalls of 15 to 20 inches depth, and as the roof in question is flat most of the snow may lie on it until melted. Will the truss shown in Fig. 162 safely hold the load mentioned, or will Mr. Kidder give the design of a similar and safer one that does not require over 4 feet rise?

Answer.—The communication with sketch of our correspondent was referred to Mr. Kidder, who furnished the following comments in reply:

The construction shown by "S. C. P." in Fig. 162 is, in the main, in accordance with scientific principles, and the truss would probably support the roof and ceiling with a light fall of snow,

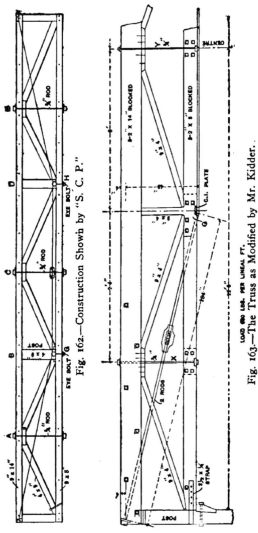

Fig. 162.—Construction Shown by "S. C. P."

Fig. 163.—The Truss as Modified by Mr. Kidder.

but under a heavy snow load the truss would sag excessively and possibly break the belly rods. By dropping the belly rods entirely below the wooden tie beam, as in Fig. 163 the stress in these rods will be greatly reduced, but even then it will require two 1⅝-inch rods if not upset or two 1½-inch rods if the ends are upset.

The construction shown in Figs. 162 and 163 consists essentially of a belly-rod truss, with the top chord or straining beam braced from the points G and H, and with a wooden tie beam to support the ceiling joists. The stress or strain in the belly rods at the ends is found by dividing the length of the rods on the slant by the rise and multiplying the quotient by the load, which is concentrated at the point G or H of Fig. 162.

In measuring the rise the distance from the center of the rods to the center of compression in the top chord should be taken. Thus in Fig. 163 an 8 x 8 inch strut would resist the compression in the straining beam, and as this beam is also reinforced to some extent by the roof, we can figure that the center of compression may be within 4 inches of the top. As the hight of the truss is not to exceed 48 inches, the distance from center of compression to center of rod cannot be over 43 inches. In Fig. 163 the writer has moved the post at G 6 inches nearer the end, so as to shorten the length of rod on the slant. The length of the belly rod on the slant is 184 inches. Dividing this by the rise, (43), we have 4.28 for the quotient. Taking the load given by "S. C. P." we find that it amounts to 677 pounds per lineal foot of truss, or say 680 pounds. The load which must be supported by the belly rods at G is that portion between rods X and Y, which is 15⅓ × 680 = 10,426 pounds. Multiplying this load by 4.28 we have 44,623 pounds as the stress in the belly rods, or 22,311 pounds on each rod. This stress will require 1⅝-inch rods if the threads are cut out of the rod or 1½-inch if the screw ends are upset.

The balance of the truss is strong enough for the loads given by "C. P. S."; 10,530 pounds for weight of roof and ceiling, however, would require a tin roof and metal or wood ceiling. If the ceiling is plastered and the roof is to be of gravel the actual

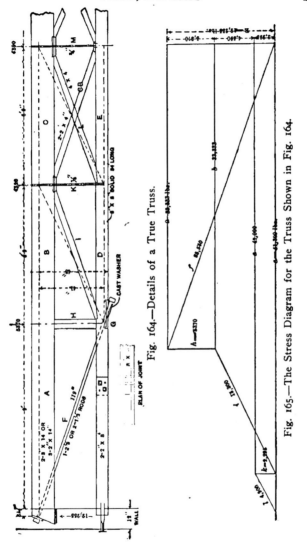

Fig. 164.—Details of a True Truss.

Fig. 165.—The Stress Diagram for the Truss Shown in Fig. 164.

weight would be about 28 pounds per square foot, or 16,380 pounds for the entire truss. The allowance for snow is equivalent to 34 pounds per square foot, which is probably ample.

In the truss shown by Fig. 162 the length of the rods on the slant is about the same as in Fig. 163, while the rise measured from center of rods is some 8 or 9 inches less; consequently the stress in the rods is increased some 25 per cent.

Fig. 164 shows another method of building a truss of the same span and hight, but in which the wooden tie beam and the braces are made to form a part of the truss. By shortening the length of the main tension rod F the ratio of length to rise is reduced to 2.9 and the stress to 35,549 pounds (2.9 × 12,255), which can be carried by one 2¼-inch rod or two 1½-inch rods. The single rod will be preferable if it can be obtained. Fig. 164 is a true truss.

Fig. 165 is the stress diagram for Fig. 164, the lines in Fig. 165 being drawn parallel to the dotted lines in Fig. 164, which represent the center lines. The length of the lines in Fig. 165 measured by the scale to which the load line is drawn gives the stress in the corresponding member of the truss; thus the line a represents the stress in A, b the stress in B, f the stress in F, i the stress in the brace I, etc.

The rod M has no stress except the weight of the ceiling. The construction shown in Fig. 164 requires a solid stick of timber 8 x 8 inches and 34 feet long for the tie beam. In the trusses Figs. 162 and 163 the tie beam can be jointed at the points G and H.

The top chord in each of the three trusses can be jointed over any one of the supports, the planks breaking joint, and all bolted together.

"S. C. P." does not show what supports the wall. If supported by frame construction the scheme shown in Fig. 163 should be used. If supported by brick walls the main bearing should be under the end of the straining beam, as in Fig. 164. The cast-iron plate at G in Fig. 163, on which the belly rods bear, should be rounded to fit the bend in the rods, and the tie beam should be blocked solid at this point, also under end of rods X and Y.

TRUSS FOR DAIRY BARN.

From W. F. C., *Elk River, Minn.*—Looking over the issue of *Carpentry and Building* for November, 1901, I noticed a communication in regard to remodeling an old dairy barn. This made me think of a barn I built in 1887, which was a structure 60 x 90 feet in size, with 20-foot posts, and as indicating the method of framing I send a blue print of one truss or bent. This form of structure was original with me, as at that time I had never seen such a truss, and the man for whom the building was being erected desired very much not to have the old-fashioned posts and braces.

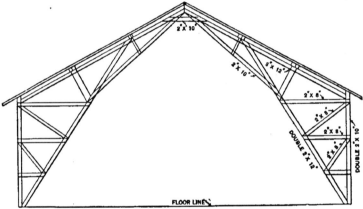

Fig. 166.—Elevation of Bent Submitted by "J. D."

An examination of the truss (Fig. 161), of which only a portion is here presented, shows by means of the dotted lines at A that a brace was put in after the frame had been raised. I found here a weak place before the braces were put in, and while the men were shingling I noticed that the roof swayed up and down, one side moving up when the opposite side went down. After the braces were put in the roof was all right. At B is an iron strap 3 inches wide, and fastened by two ⅝-inch bolts. This barn is standing to-day (April, 1902) in just as good shape as when it was erected, and I can see it now from where I am writing this letter.

BENTS FOR 12-INCH PLANK FRAME BARN.*

From J. D., *Ubly, Mich.*—I send herewith a rough sketch, Fig. 166, showing one of the bents of a 12-sided plank frame barn, and would like to have the architectural readers of the paper state whether or not it will be strong enough for the purpose, and, if not, wherein changes can be made to advantage.

Note.—With a view to obtaining the opinion of an expert who has had long experience in the construction of plank frame barns, we submitted the inquiry of our correspondent above to John L. Shawver, who furnishes the following in reply:

In the first place, the purlin posts in the sketch of the correspondent are so sloping that while they brace well they are not in position to sustain the most weight, and these with the roof supports are so long that they have to be spliced. While this is easily done in the case of posts, it is not so readily performed with supports, and, when so done, it will not present as attractive an appearance as would otherwise be the case.

In the second place, one of the weakest points about the barn is the shape. It is true it would be a novelty in most communities, but, like the round barns, is wasteful of material. It is out of the question to place joists, rafters, flooring, sheathing, roofing, etc., on barns of this shape without much waste of both materials and labor. Then, too, it is next to impossible to provide for satisfactory lighting or ventilation, both of which are essential features in every up-to-date barn.

The frame construction indicated in Fig. 167 is stronger, and at the same time gives more open space in the interior, this being secured by running the purlin posts up to the first purlin plate, instead of to the second, and supporting the second plate on the truss brace. This saves in the lengths of the purlin posts, but requires the same length of supports and longer sub-supports.

The form of construction indicated in Fig. 168 is, in my opinion, preferable to either of the others, if it is found that the vertical

---

*Reproduced from *Plank Frame Barn Construction*, published by the David Williams Company, New York.

posts set in 10 feet will not in any way interfere with the purpose
of the structure.   In this case all the timbers are either shortened

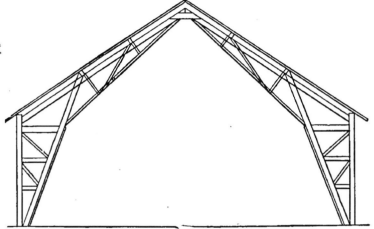

Fig. 167.—One Form of Plank Frame Suggested by Mr. Shawver.

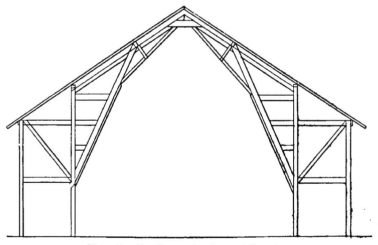

Fig. 168.—The Preferable Style of Framing.

or placed in such shape that they may be spliced without in any way weakening the structure. Whichever form may be used by the correspondent, it is important in bents or arches of this size that the purlin posts should be braced on the inner edge with 2 x 6, which will add materially to the strength of the frame, and at the same time prevent any tendency to spring sideways either in the raising or from the pressure of the hay or grain within after the building is completed.

INDEX

222INDEX

PAGE

Trestles, bracing of.. 171
Triangular trusses, action of the stresses in....................... 52
Triangular trusses, stress diagram for, example................... 92
Triangular trusses, types of....................................... 50
Trimmers, to find the proper dimensions of....................... 4
Trimmers, to find the strength of................................. 29
Truss, definition of.. 49
Truss, diagrams, how lettered..................................... 90
Truss, joints, examples of..................................119, 159
Truss, reactions.. 90
Truss, supported at the center.................................... 203
Trussed floors.. 45
Trusses, for flat roofs... 57
Trusses, lattice.. 70
Trusses, proportioning the joints of.............................. 118
Trusses, proportioning the members to the stresses............... 107
Trusses, scissors... 75
Trusses, Types of... 48

Washers, dimensions of.. 143
Washers, how to proportion to stress in the rod................... 140
Weight of, ceilings... 85
Weight of, floor joists, table.................................... 25
Weight of, mill shavings, per cubic foot.......................... 177
Weight of, partitions... 32
Weight of, purlins, per square foot of roof....................... 83
Weight of, rafters, per square foot of roof....................... 83
Weight of, roofing materials...................................... 84
Weight of, roof trusses, per square foot of roof, table........... 83
Weight of, snow on roofs.. 84
Weight of, wooden floor construction.............................. 24
Wind-bracing, of buildings. 162; example of...................... 162
Wind-bracing, of frame factories................................. 170
Wind-bracing, of framed spires................................... 177
Wind pressure, against tall buildings............................. 164
Wind pressure, allowance for on roofs............................. 85
Wind pressure, resistance of buildings to......................... 169
Wind stresses, in a trestle....................................... 171

THE END.

Lightning Source UK Ltd.
Milton Keynes UK
13 October 2009

144931UK00001B/25/P